Rapid Math in 10 Days

RAPID
Math in 10 Days

The Quick-and-Easy Program for Mastering Numbers

Edward H. Julius

A PERIGEE BOOK

To Marion, Marina, and Alexandra

I would like to gratefully acknowledge Dolores McMullan for her meticulous and thoughtful editing of the manuscript. I would also like to thank Steve Ross for once again believing in me.

A Perigee Book
Published by The Berkley Publishing Group
200 Madison Avenue
New York, NY 10016

Copyright © 1994 by Edward H. Julius.

Book design by Irving Perkins Associates, Inc.

Cover design by Bob Silverman, Inc.

First Perigee edition: September 1994

Published simultaneously in Canada.

Library of Congress Cataloging-in-Publication Data

Julius, Edward H., 1952–
    Rapid math in ten days : the quick-and-easy program for mastering
numbers / Edward H. Julius.—1st Perigee ed.
      p.   cm.
    "A Perigee book."
    ISBN 0-399-52129-1
    1. Mental arithmetic. I. Title.
QA107.J85 1994
513.9—dc20
                             94-5998
                             CIP

Printed in the United States of America

10 9 8 7 6 5 4 3 2 1

CONTENTS

Throughout the book, a number of "Number Curiosities," tips, and anecdotes are provided for your enlightenment and enjoyment.

INTRODUCTION

"Man is still the most extraordinary computer of all."
—JOHN F. KENNEDY

I would wholeheartedly agree with the late President's dictum, despite his inadvertent sexism. Unfortunately, mental math—perhaps the most efficient use of our neurological computer—is quickly becoming a lost art. Mental math is the performance of a calculation without the use of either a calculator or pencil and paper, and many think it's an outmoded skill. Why should *you* develop mental math skills in this "age of the calculator"? There are several reasons:

- Sometimes it's awkward to use a calculator, as when your hands are full, or when you're driving.
- Other times it's embarrassing to use a calculator, as when the calculation is fairly simple, such as dividing the check in a restaurant.
- Think of all the times you'd rather perform a calculation *secretly*, as when you feel, for instance, that a store clerk has shortchanged you, but you'd rather he or she not know that you're double-checking the math.
- Occasionally, it's a hindrance to use a calculator, as when you're engaged in rapid-fire negotiations, and must manipulate numbers almost instantly.

- Sometimes your calculator's batteries die at inconvenient moments. I can't tell you how many of my college students over the years have absolutely panicked when their calculators have suddenly died on them. And when I suggest that they perform the math with pencil and paper (never mind performing it in their heads), they usually stare at me in disbelief.
- Finally, when you become adept at mental math, you'll often find that using a calculator actually slows you down. Really!

Don't get me wrong. I think the calculator is a marvel of modern technology, and I overwhelmingly prefer it to the slide rule I grew up using and hating. However, I refuse to be a slave to my calculator, and I use it only when the calculations are very long and cumbersome.

Convinced? Then let's proceed.

This book contains forty carefully chosen rapid calculation techniques, organized into a ten-day program. Each day, you will learn four techniques which build upon themselves, beginning with addition and subtraction, through multiplication and division and ending with some very clever estimation techniques.

For each technique, two practical applications have been provided, as well as three demonstration problems and step-by-step explanations. When you feel you understand the basic strategy, then proceed to the exercise section. First, there are "basic exercises" to test your skills with relatively easy calculations. At a minimum, you should perform all twelve or thirteen basic exercises within a section. However, if you would like to open yourself up to a much wider world of possibilities and don't mind stretching your mental powers a bit, try the "Math Master" exercises as well.

Don't overlook the "Food for Thought" sections. They provide special insight into the rapid math technique at hand, and will teach you more about how to think about and understand numbers.

At the end of each day (four techniques later), you will take a brief review quiz. To test your progress after five days, a midterm exam is provided, and to test your abilities at the program's conclusion, you will take a 50-question final exam. Don't let the word "exam" unnerve you; think of these as focused practice sessions that allow you to check—and take pride in—your progress. In all, this program contains over 1,000 practice problems for you to work. Remember, the best way to learn is by doing.

Throughout the book, you'll also find mathematical diversions of sorts. These "number curiosities," "quick questions," and anecdotes are provided for your enlightenment and amusement. My real agenda in writing this book is to get people to think about and truly enjoy numbers. Let me share an alarming story with you. Recently, I went to my local delicatessen to buy ¾ of a pound of meat. When the deli employee had placed ½ a pound on the scale, he looked puzzled, and asked me, "Is ¾ more or less than ½?" Although many people are intimidated by numbers, anyone can develop a basic understanding of them. And with a bit of practice, numbers can not only be mastered, they can be fun!

You might not have the time and inclination to learn every technique in this program. Or you may need more than ten days to complete it, if you feel you're overloading yourself. One of the advantages of a book like this is that it enables you to tailor the program to your strengths and preferences.

After completing the program, pick the ten or twenty techniques that you like best, or think are most useful to you, and simply set aside the others. Develop a balanced repertoire of techniques, and practice them as much as you possibly can. Frequently, you'll find that you can use two techniques at once or alternative techniques to solve problems. So the more techniques you master, the more likely you will be able to reduce any calculation to child's play.

But before you begin, I urge you to review the concepts, terminology and other material covered in the next few pages. A thorough understanding of this information is imperative for success in this program.

When you complete the program, you'll relish your newfound mathematical powers. You will attack numbers with confidence, accuracy and lightning speed. In short, you will be a mental-math wizard. All it takes is learning the tricks of the trade!

—ED JULIUS

12 ABSOLUTELY ESSENTIAL MATH CONCEPTS FOR YOU TO REVIEW

1. You should completely understand the concept of "place value," or the significance of each digit within a number. For example, in the number 752, the "7" represents 700, the "5" represents 50, and the

"2" represents 2. So 752 can be viewed as $700 + 50 + 2$. Place value also applies to digits after the decimal point. For example, in the number 3.14, the "1" represents $\frac{1}{10}$ and the "4" represents $\frac{4}{100}$.

2. You should understand when you can affix a zero to a number without altering its value. For example, the number 86 can also be expressed as 086 (though it usually isn't), or as 86.0, 86.00, etc. It should be obvious that 860 and 806 would not be equivalent to 86.

3. Any number of numbers can be added in any order to produce the same sum. For example, $17 + 46 = 46 + 17$.

4. Any number of numbers can be multiplied in any order to produce the same product. For example, $23 \times 11 = 11 \times 23$.

5. Subtraction is the inverse of addition. For example, if $7 + 9 = 16$, then $16 - 9 = 7$, and $16 - 7 = 9$.

6. Division is the inverse of multiplication. For example, if $8 \times 7 = 56$, then $56 \div 7 = 8$, and $56 \div 8 = 7$.

7. Division can be represented in three different ways. For example, 75 divided by 4 could be written as $75 \div 4$, $\frac{75}{4}$, or $4\overline{)75}$. Similarly, addition, subtraction, and multiplication calculations can be expressed either horizontally or vertically. Most of the calculations in this book are expressed horizontally to force you to perform the computation the "Excell-erated" way.

8. Squaring a number is the same as multiplying the number by itself. To square the number 14, for example, you would write 14^2, or 14×14.

9. A percentage is a quantity relative to 100. For example, 30% is the same as $\frac{30}{100}$ (or $\frac{3}{10}$), and 7% is the same as $\frac{7}{100}$.

10. Percentages, decimals, and fractions have a certain interrelationship. For example, 80% can also be expressed as 0.8 or as $\frac{80}{100}$ (or $\frac{8}{10}$). Similarly, the number $3\frac{1}{2}$ can also be expressed as 3.5.

11. The term "power of 10" refers to the number of times 10 appears in a multiplication by itself. For example, 10^2 (spoken as "ten to the second power" or simply "ten squared") means 10×10, or 100. Similarly, 10^5 (spoken as "ten to the fifth power") means $10 \times 10 \times 10 \times 10 \times 10$, or 100,000. To multiply a number by a power of 10, simply affix the appropriate number of zeroes. So $73 \times 10 = 730$, $73 \times 100 = 7,300$, and $73 \times 1,000 = 73,000$.

12. To divide a number by a power of 10, simply move the decimal point the appropriate number of places to the left or eliminate the appropriate number of right-hand zeroes. So $924 \div 10 = 92.4$, and $924 \div 100 = 9.24$. Similarly, $3,600 \div 10 = 360$, and $3,600 \div 100 = 36$.

Taken one step further, when both the dividend and divisor contain right-hand zeroes, you can cancel an equal number of zeroes to simplify the computation. For example, 40,000 ÷ 300 could be expressed as 4,000 ÷ 30 or as 400 ÷ 3.

SOME IMPORTANT TERMINOLOGY

```
 47 ← Addend          Multiplicand →  15 ← Factor
+36 ← Addend          Multiplier  → ×42 ← Factor
 83 ← Sum             Product     → 630
```

```
131 ← Minuend         3,600  ÷  500  =  7.2
−89 ← Subtrahend        ↑        ↑        ↑
 42 ← Difference      Dividend  Divisor  Quotient
```

ADDITION/SUBTRACTION TABLE

+	1	2	3	4	5	6	7	8	9	10	11	12
1	2	3	4	5	6	7	8	9	10	11	12	13
2	3	4	5	6	7	8	9	10	11	12	13	14
3	4	5	6	7	8	9	10	11	12	13	14	15
4	5	6	7	8	9	10	11	12	13	14	15	16
5	6	7	8	9	10	11	12	13	14	15	16	17
6	7	8	9	10	11	12	13	14	15	16	17	18
7	8	9	10	11	12	13	14	15	16	17	18	19
8	9	10	11	12	13	14	15	16	17	18	19	20
9	10	11	12	13	14	15	16	17	18	19	20	21
10	11	12	13	14	15	16	17	18	19	20	21	22
11	12	13	14	15	16	17	18	19	20	21	22	23
12	13	14	15	16	17	18	19	20	21	22	23	24

Any number added to zero equals that number. To use this table to subtract, find the minuend in the body of the table and read up or to the left to determine the subtrahend and difference. For example, 15 − 9 = 6.

MULTIPLICATION/DIVISION TABLE

×	1	2	3	4	5	6	7	8	9	10	11	12
1	1	2	3	4	5	6	7	8	9	10	11	12
2	2	4	6	8	10	12	14	16	18	20	22	24
3	3	6	9	12	15	18	21	24	27	30	33	36
4	4	8	12	16	20	24	28	32	36	40	44	48
5	5	10	15	20	25	30	35	40	45	50	55	60
6	6	12	18	24	30	36	42	48	54	60	66	72
7	7	14	21	28	35	42	49	56	63	70	77	84
8	8	16	24	32	40	48	56	64	72	80	88	96
9	9	18	27	36	45	54	63	72	81	90	99	108
10	10	20	30	40	50	60	70	80	90	100	110	120
11	11	22	33	44	55	66	77	88	99	110	121	132
12	12	24	36	48	60	72	84	96	108	120	132	144

Any number multiplied by zero equals zero. To use this table to divide, find the dividend in the body of the table and read up or to the left to determine the divisor and quotient. For example, $54 \div 6 = 9$.

SOME SQUARES YOU SHOULD KNOW

$1^2 = 1$	$6^2 = 36$	$11^2 = 121$	$16^2 = 256$
$2^2 = 4$	$7^2 = 49$	$12^2 = 144$	$17^2 = 289$
$3^2 = 9$	$8^2 = 64$	$13^2 = 169$	$18^2 = 324$
$4^2 = 16$	$9^2 = 81$	$14^2 = 196$	$19^2 = 361$
$5^2 = 25$	$10^2 = 100$	$15^2 = 225$	$20^2 = 400$

TABLE OF EQUIVALENCIES

$1/100 = 0.01\ \ = 1\%$ $3/8 = 0.375 = 37\frac{1}{2}\%$

$1/50\ \ = 0.02\ \ = 2\%$ $2/5 = 0.4\ \ \ = 40\%$

$1/40\ \ = 0.025 = 2\frac{1}{2}\%$ $1/2 = 0.5\ \ \ = 50\%$

$1/25\ \ = 0.04\ \ = 4\%$ $3/5 = 0.6\ \ \ = 60\%$

$1/20\ \ = 0.05\ \ = 5\%$ $5/8 = 0.625 = 62\frac{1}{2}\%$

$1/10\ \ = 0.1\ \ \ = 10\%$ $2/3 = 0.6\overline{66} = 66\frac{2}{3}\%$

$1/9\ \ \ = 0.11\overline{1} = 11\frac{1}{9}\%$ $3/4 = 0.75\ \ = 75\%$

$1/8\ \ \ = 0.125 = 12\frac{1}{2}\%$ $4/5 = 0.8\ \ \ = 80\%$

$1/5\ \ \ = 0.2\ \ \ = 20\%$ $7/8 = 0.875 = 87\frac{1}{2}\%$

$1/4\ \ \ = 0.25\ \ = 25\%$ $x/x = 1.0\ \ \ = 100\%$

$1/3\ \ \ = 0.33\overline{3} = 33\frac{1}{3}\%$

FINAL NOTE BEFORE YOU GET STARTED

When performing a calculation, always give a quick test to your solution to determine if it's "in the ballpark." Let's say you've been asked to multiply 1.5 by 30, and accidentally arrive at an answer of 450. By simple inspection, you should know that the correct solution is somewhere between 30 (1 × 30) and 60 (2 × 30), since 1.5 is between 1 and 2. You can then lop the zero off the answer of 450, to arrive at the correct answer, 45. Performing this quick test takes practice, and sometimes imagination, but is invaluable in mastering the art of rapid calculation.

DAY 1:

ADDITION

DAY 1

TECHNIQUE 1:
Adding by Grouping and Reordering

BASIC APPLICATION

Let's begin at the supermarket. You've just placed seven items in your shopping cart and would like to know how much you've spent so far. To obtain a quick estimate, you round each item up or down to the nearest dollar (a very powerful technique in itself), and arrive at the following amounts: $6, $2, $2, $7, $5, $3, and $4. Based on these rounded amounts, how much have you spent?

Calculation: $6 + 2 + 2 + 7 + 5 + 3 + 4 = ?$

The Old-Fashioned Way

$$\begin{array}{r} 6 \\ 2 \\ 2 \\ 7 \\ 5 \\ 3 \\ + 4 \\ \hline 29 \end{array}$$

The numbers are added slowly, in order, and one at a time.

The Excell-erated Way

Although many people use speed-reading techniques for the printed word, very few use similar methods for a column of numbers. To add the numbers listed to the left, they might think to themselves, "6 plus 2 equals 8, 8 plus 2 equals 10, 10 plus 7 equals 17," and so on. Instead, we're going to recommend three techniques that will cut your addition time substantially: (A) *visualize two or three numbers as their sum*, (B) *add out of*

3

order, where appropriate, and (C) *add in multiples of 10, wherever possible.* Let's look below to see what all this means.

A. VISUALIZE TWO OR THREE NUMBERS AS THEIR SUM:

When looking at the first two numbers listed on page 3, don't say to yourself, "6 plus 2 equals 8." Instead, look at the 6 and the 2 together, and instantly visualize "8." (With practice, you'll be able to look at the 6 and both of the 2's, and immediately visualize "10.") Then focus on the second 2 and the 7, and immediately visualize "9." Remembering that you have a running total of 8 to this point, visualize the sum of 8 and 9 as 17. Continue adding two at a time, or just one if a combination of two are cumbersome, until you obtain the sum.

B. ADD OUT OF ORDER, WHERE APPROPRIATE:

If you see that you have a simpler addition at hand don't hesitate to skip ahead a little when adding. Technique "C" below relies heavily upon skipping ahead, though this method is not limited to achieving multiples of 10. Determining when to skip is a matter of personal preference, and does take practice. However, don't skip ahead *too* far, and remember to come back to those numbers you've passed over.

C. ADD IN MULTIPLES OF 10, WHEREVER POSSIBLE:

One of the most powerful mental math techniques you can use, whether you are adding, subtracting, multiplying, dividing, or squaring, is to achieve a multiple or power of 10 whenever possible (even if you have to "cheat" to do it). We do this, quite simply, because it's extremely easy to deal with numbers like 10, 40, 100, and 300. In our supermarket example above, let's say you've already visualized the first three numbers (6, 2 and 2) as "10." You can then see that the 7 and (skipping ahead) the 3 total 10. Now you're at 20. Finally, you visualize the 5 and the 4 (out of order, as they are) as "9," and you've got 29 as the answer. With practice, you will be able to look at those seven numbers and think, "10, 20, 29" in almost no time at all.

Let's Try One More Basic Calculation: $5 + 7 + 6 + 9 + 4 + 2 + 1$

Using our three strategies, we visualize the 5 and the 7 as "12." Then we look at the 6 and immediately think, "18." (At this point, *don't* say to yourself, "12 plus 6 equals 18." It uses up too much time.) Hopefully you've just spotted a "10 combination"—the 9 and the 1. You're now up to 28, and must remember to backtrack and pick up the 4 and the 2. It's probably best to add the 2 first, because you would then reach a nice even 30. Finally, the 4 will bring you to the answer, 34. Although this explanation was lengthy, our recommended way to add these numbers is to think "12, 18, 28, 30, 34," in no more than three seconds (with practice, of course!).

Math-Master Application

You've just completed nine holes of golf, and your scorecard looks like this: 7 7 5 5 3 6 9 2 4. What is your nine-hole total?

Calculation: $7 + 7 + 5 + 5 + 3 + 6 + 9 + 2 + 4 = ?$

Math-Master Solution

There are, of course, a number of ways to approach this calculation, using our three strategies. Even adding successive pairs would work— you would think, "14, 24, 33, 44, 48." Try skipping ahead or locating combinations of 10 to arrive at the answer, 48.

Food for Thought

A further application of this technique is to multiply when you have recurring numbers. For example, $8 + 7 + 8 + 8 + 6$ can be added quickly by spotting the three 8's and thinking "24" (that is, 3×8), to begin the calculation. Then the 6 will bring you to a nice even 30, and backtracking to the 7 will produce the answer, 37. So you would think, "24, 30, 37" in almost no time.

Once you've learned the basic strategies for adding, the key is to *practice* and *experiment* to see which combination of techniques works best for you.

NOW IT'S YOUR TURN

Basic Exercises

1. $3 + 3 + 3 + 7 + 4 + 1 + 9 =$
2. $2 + 8 + 5 + 4 + 7 + 1 + 6 =$
3. $9 + 9 + 7 + 5 + 1 + 4 + 8 =$
4. $4 + 3 + 3 + 8 + 6 + 2 + 8 =$
5. $7 + 7 + 9 + 7 + 1 + 4 + 4 =$
6. $5 + 8 + 2 + 3 + 6 + 6 + 7 =$
7. $1 + 5 + 7 + 4 + 3 + 9 + 2 =$
8. $6 + 4 + 9 + 9 + 3 + 5 + 7 =$
9. $8 + 7 + 2 + 5 + 4 + 1 + 6 =$
10. $2 + 2 + 6 + 9 + 5 + 4 + 8 =$
11. $2 + 8 + 7 + 6 + 5 + 4 + 3 =$
12. $6 + 3 + 6 + 6 + 9 + 5 + 8 =$
13. $1 + 3 + 5 + 7 + 6 + 4 + 2 =$

Math Masters

14. $8 + 3 + 4 + 7 + 1 + 1 + 6 + 9 + 2 =$
15. $5 + 4 + 7 + 1 + 8 + 8 + 6 + 2 + 3 =$
16. $3 + 9 + 9 + 9 + 5 + 7 + 8 + 4 + 6 =$
17. $7 + 4 + 4 + 9 + 1 + 3 + 6 + 3 + 8 =$
18. $1 + 5 + 6 + 3 + 9 + 7 + 9 + 2 + 2 =$
19. $6 + 8 + 5 + 9 + 4 + 2 + 7 + 7 + 1 =$
20. $9 + 7 + 5 + 3 + 1 + 8 + 6 + 4 + 2 =$

(*See answers on page 191*)

DAY 1

TECHNIQUE 2:
Adding from Left to Right I

BASIC APPLICATION

You've entered the teaching profession, and are grading your first exam. The exam is six pages long, and you've computed the point total for everyone for each of the six pages. Now it's time to add up the page totals to determine the numerical grade for each student. The page totals for the first exam are 18, 23, 9, 11, 27, and 6. What is the student's grade?

Calculation: $18 + 23 + 9 + 11 + 27 + 6 = ?$

The Old-Fashioned Way

$$
\begin{array}{r}
3 \\
18 \\
23 \\
9 \\
11 \\
27 \\
+\ 6 \\
\hline
94
\end{array}
$$

First the ones column is totaled, the 3 is carried, and then the tens column is totaled.

The Excell-erated Way

In our exam example, you actually couldn't use the old-fashioned way, because the numbers to be added are on separate pages, and not in columnar form. Anyway, the best way to add numbers that are presented one at a time (as in the example above) is to first add the tens digit and then add the ones digit. Let's try it and see how our first student scored on the exam.

7

The Excell-erated Solution

Remember that a "1" in the tens place really represents 10, a "2" represents 20, and so on. So every time you add a tens digit, make sure you add the appropriate multiple of 10. In any case, let's reproduce the calculation in the space below, and apply this technique.

18→   We begin by just saying or thinking, "18."
23→   Then say, "28, 38, 41." (Add 10, 10 again, and then 3.)
 9→   Then say, "50." (Just add the 9 to the 41.)
11→   Then say, "61." (It's easy just to add the whole number here.)
27→   Then say, "71, 81, 88." (Add 10, 10 again, and then 7.)
+ 6→   Then say, "94," which completes the calculation.

(NOTE: For the numbers 23 and 27 above, you don't have to add 10, and then 10 again. You could simply add 20, though it might actually be slower to do it that way.)

Let's Try One More Basic
Calculation: 27 + 18 + 5 + 34 + 12 + 46

27→   Think or say, "27."
18→   Then say, "37, 45."
 5→   Then say, "50."
34→   Then say, "84." (We can add the entire number here.)
12→   Then say, "94, 96." (Or add the entire number at once.)
+ 46→ Then say, "106, 116, 126, 136, 142." (142 is the sum.)

(NOTE: When adding the last number, you can skip directly from 96 to 136.)

Math-Master Application

You've just purchased five items of clothing that carry price tags of $61, $27, $52, $8, and $38. How much have you spent? (This calculation requires more concentration than the basic problems above because the numbers are generally larger.)

Calculation: 61 + 27 + 52 + 8 + 38 = ?

Math-Master Solution

61→   Think or say, "61."

27→   Then say, "81, 88." (Or count by tens, if you prefer.)

52→   Then say, "138, 140." (Or count by tens, if you prefer.)

8→   Then say, "148."

+ 38→   Then say, "178, 186." (Or count by tens, if you prefer.)
      The sum is 186.

Food for Thought

Technique #2 can apply to numbers of any size. For example, you could add numbers in the hundreds, by first adding the hundreds digit of a number, then the tens digit, and finally the ones digit. Generally speaking, the larger the numbers and the longer the list, the more this technique is going to require your concentration and skill.

NOW IT'S YOUR TURN

Basic Exercises

1. $17 + 41 + 3 + 22 + 35 + 11 =$

2. $9 + 37 + 28 + 13 + 6 + 41 =$

3. $24 + 15 + 40 + 8 + 32 + 19 =$

4. $47 + 3 + 36 + 14 + 21 + 21 =$

5. $5 + 34 + 45 + 28 + 14 + 7 =$

6. $16 + 25 + 2 + 39 + 42 + 7 =$

7. $33 + 8 + 27 + 13 + 48 + 10 =$

8. $7 + 9 + 49 + 26 + 35 + 18 =$

9. $27 + 22 + 1 + 44 + 38 + 19 =$

10. $30 + 18 + 42 + 6 + 35 + 24 =$

11. $13 + 34 + 29 + 7 + 18 + 46 =$

12. $9 + 40 + 28 + 31 + 12 + 49 =$

13. $46 + 37 + 29 + 5 + 14 + 33 =$

Math Masters

14.	$15 + 49 + 77 + 20 + 51 =$
15.	$61 + 32 + 54 + 80 + 19 =$
16.	$66 + 37 + 8 + 72 + 21 =$
17.	$46 + 20 + 99 + 14 + 57 =$
18.	$58 + 31 + 72 + 6 + 19 =$
19.	$88 + 12 + 32 + 67 + 24 =$
20.	$99 + 77 + 55 + 33 + 11 =$

(See answers on page 191)

TOTALING TIP

Adding long columns of large numbers is not only cumbersome, but there is a pretty good likelihood that you'll make an error somewhere along the way. Accordingly, why not divide the column of numbers into sections, add within each section, and then add the section totals. For example, if you're adding fifty large numbers, why not add the first ten numbers, then the next ten, and so on. This way, not only can you locate errors more easily, but you won't have the task of accumulating such large sums in your head (and probably losing count).

DAY 1

TECHNIQUE 3:
Adding from Left to Right II

BASIC APPLICATION

Because this technique is a variation on Technique #2, we will assume the same set of facts and numbers, for purposes of easy comparison. You may recall that we were adding exam points of 18, 23, 9, 11, 27, and 6.

Calculation: $18 + 23 + 9 + 11 + 27 + 6 = ?$

The Old-Fashioned Way

```
  3
 18
 23
  9
 11
 27
+ 6
───
 94
```

The Excell-erated Way

To apply this technique, we will still be adding from left to right. However, we will first be adding all the tens digits, and then all the ones digits. This technique is probably more efficient than Technique #2 when the numbers are placed in columnar form. Let's see how this variation on Technique #2 works.

The Excell-erated Solution

Once again, remember that each number in the tens place represents a multiple of ten. So a 7, for example, in the tens column really represents 70.

18	Begin at the top of the tens column and read down,
23	counting "10, 30, 40, 60." Then proceed down the ones
9	column, continuing to count, "68, 71, 80, 81, 88, 94." To
11	save a split-second, you can proceed from the *bottom* of
27	the ones column and work your way up.
+ 6	

NOTE: When adding these columns of numbers, don't hesitate to use Technique #1 (grouping, reordering, and finding combinations of 10) to perform the calculation even faster!

Let's Try One More Basic
Calculation: 27 + 18 + 5 + 34 + 12 + 46

27	Begin at the top of the tens column and read down,
18	counting, "20, 30, 60, 70, 110." Then proceed down (or
5	up) the ones column, continuing to count, "117, 125,
34	130, 134, 136, 142" (or "116, 118, 122, 127, 135, 142").
12	
+ 46	

Math-Master Application

Again, we will assume the same set of facts as in Technique #2 for purposes of comparison. You may recall that you've purchased items of clothing costing $61, $27, $52, $8, and $38. Using Technique #3 this time, compute the total amount spent.

Calculation: 61 + 27 + 52 + 8 + 38 = ?

Math-Master Solution

```
  61
  27
  52
   8
+ 38
─────
```

Begin at the top of the tens column and read down, counting, "60, 80, 130, 160." This time we'll proceed up the ones column, because it is slightly faster. Continue counting, "168, 176, 178, 185, 186." The sum is 186.

Food for Thought

As is true with Technique #2, Technique #3 can apply to numbers of any size. Just remember what each place value represents. These two techniques are superior to the conventional method of adding because (a) there is no need to carry, and (b) you obtain the answer all at once, rather than in portions. Thus, there is no writing until you've obtained the answer.

NOW IT'S YOUR TURN

Basic Exercises

1.	2.	3.	4.
35	6	36	25
40	14	51	49
11	37	5	7
4	18	27	12
17	23	18	50
+ 28	+ 9	+ 2	+ 33

5.	6.	7.	8.
25	42	1	13
34	8	21	4
7	33	39	8
48	27	40	27
2	10	13	40
+ 19	+ 9	+ 6	+ 16

9.	3	10.	17	11.	28	12.	7
	16		29		37		42
	47		9		41		33
	38		43		2		25
	22		31		15		14
	+ 5		+ 6		+ 26		+ 3

Math Masters

13.	41	14.	56	15.	84	16.	38
	38		42		25		47
	26		1		37		9
	2		98		4		66
	+ 77		+ 25		+ 12		+ 13

17.	54	18.	6	19.	27	20.	11
	21		88		50		22
	8		47		14		44
	92		19		64		77
	+ 33		+ 30		+ 3		+ 99

(See answers on page 191)

DAY 1

TECHNIQUE 4:
Adding by Altering

BASIC APPLICATION

We finish Day 1 with another powerful technique. Let's assume you're playing Scrabble, and you currently have 73 points. Spotting a place on the board to use your "Q," you score 49 points. What is your score now?

Calculation: 73 + 49 = ?

The Old-Fashioned Way

$$\begin{array}{r} \overset{1}{73} \\ +\ 49 \\ \hline 122 \end{array}$$

The Excell-erated Way

Most people who play Scrabble do add points the conventional way, rather than in their head. As you'll see as you proceed through this program, there are several altering techniques. This one involves converting one of the numbers into a more "user-friendly" number. In our Scrabble example, we will convert the 49 into a 50, add 73 and 50, and then subtract 1. Remember that it's always easier to add a number that is a multiple of 10 than one that is not.

15

To add 73 and 49, answer the following questions

A. Which number is closer to a multiple of   Answer: 49
    10?
B. What multiple of 10 is 49 close to?   Answer: 50
C. What is 73 + 50?   Answer: 123
D. How much was added to 49 to obtain   Answer: 1
    50?
E. What is 123 − 1?   Answer: 122 (solution)

(SUMMARY: We add 1 initially, but then subtract it as the last step.)

Let's Try One More Basic Calculation: 38 + 56

To add 38 and 56, answer the following questions

A. Which number is closer to a multiple of   Answer: 38
    10?
B. What multiple of 10 is 38 close to?   Answer: 40
C. What is 40 + 56?   Answer: 96
D. How much was added to 38 to obtain   Answer: 2
    40?
E. What is 96 − 2?   Answer: 94 (solution)

Math-Master Application

You currently have $189 in the bank, and are about to deposit $344 more. What will your new balance be?

Calculation: $189 + $344 = ?

To add $189 and $344, answer the following questions
(For these larger numbers, let's try altering to the closest multiple of 100.)

A. Which number is closer to a multiple of   Answer: 189
    100?
B. What multiple of 100 is 189 close to?   Answer: 200
C. What is 200 + 344?   Answer: 544
D. How much was added to 189 to obtain   Answer: 11
    200?
E. What is 544 − 11?   Answer: 533 (solution)

Food for Thought

As you may have guessed, this technique also applies to multiples of 1,000, 10,000, and so on. Also note that the number being altered must be close to a multiple of 10, etc., and usually should end in an 8 or a 9. You can apply the altering technique any way you like so long as it truly speeds up the calculation. For instance, in our initial example, you could simultaneously add 1 to the 49 and subtract 1 from the 73, converting the calculation to a simpler $50 + 72$.

NOW IT'S YOUR TURN

Basic Exercises

1. $18 + 94 =$
2. $59 + 64 =$
3. $45 + 78 =$
4. $97 + 29 =$
5. $48 + 84 =$
6. $75 + 39 =$
7. $66 + 98 =$
8. $58 + 17 =$
9. $79 + 34 =$
10. $93 + 38 =$
11. $75 + 19 =$
12. $68 + 55 =$
13. $28 + 57 =$

Math Masters

14. $79 + 145 =$
15. $134 + 38 =$
16. $113 + 59 =$
17. $68 + 127 =$
18. $189 + 46 =$
19. $129 + 55 =$
20. $64 + 178 =$

(See answers on page 191)

TOTALING TIP

When adding, do you hate to carry? Well, it really isn't necessary at all. Take a look at the example to the right. You'll note that the sum of the ones digits is 21, which is written first. Then the sum of the tens digits, 24, is written below that, one place to the left. Add, and you've got the answer. Want to reverse the procedure and add the tens digits first (but remembering to place the sum in the proper columns)? This technique will still work. Try it!

```
   89
   74
 + 98
 ----
   21
   24
 ----
  261
```

NUMBER CURIOSITY

$$1,371,742 \times 9 = 12,345,678$$
$$13,717,421 \times 9 = 123,456,789$$

DAY 1 REVIEW QUIZ

DIRECTIONS: Work each exercise as quickly as possible, using the techniques just learned.

Basic Exercises

1. $5 + 2 + 3 + 7 + 4 + 9 + 1 =$

2. $18 + 35 + 6 + 22 + 49 + 10 =$

3. $69 + 55 =$

4. $9 + 4 + 6 + 1 + 5 + 8 + 7 =$

5. $33 + 8 + 14 + 40 + 27 + 19 =$

6. $96 + 28 =$

7.
$$\begin{array}{r} 26 \\ 15 \\ 9 \\ 41 \\ 28 \\ + 37 \\ \hline \end{array}$$

8.
$$\begin{array}{r} 5 \\ 22 \\ 38 \\ 16 \\ 3 \\ + 47 \\ \hline \end{array}$$

Math Masters

9. $8 + 4 + 5 + 7 + 1 + 3 + 9 + 6 + 2 =$

10. $27 + 64 + 30 + 81 + 49 =$

11. $46 + 179 =$

12.
$$\begin{array}{r} 25 \\ 98 \\ 4 \\ 46 \\ + 52 \\ \hline \end{array}$$

(See answers on page 191)

DAY 2:

SUBTRACTION

DAY 2

TECHNIQUE 5:
Subtracting by Adding I

BASIC APPLICATION

An automobile dealership begins the day with 81 vehicles. By day's end, 57 remain unsold. How many vehicles were sold that day?

Calculation: $81 - 57 = ?$

The Old-Fashioned Way

$$
\begin{array}{r}
\overset{7}{\cancel{8}}\overset{11}{\cancel{1}} \\
-\ 57 \\
\hline
24
\end{array}
$$

We begin with the ones digit, must "borrow," and finish with the tens digit

The Excell-erated Way

Not only will we subtract by adding, we will subtract from left to right, all in one motion. The theory behind this technique is that it is normally easier to add than to subtract. Using our automobile example, we work backwards, asking ourselves, "57 plus what equals 81?" Let's see exactly how this very useful technique works.

To subtract 57 from 81, answer the following questions

A. 57 plus what multiple of 10 will bring us to 81 or just below?

Answer: Adding 20 will bring us to 77.

B.   77 plus what equals 81?   Answer: 4
C.   What is 20 + 4?   Answer: 24 (the solution)

Let's Try One More Basic Calculation: 85 − 36

To subtract 36 from 85, answer the following questions

A.   36 plus what multiple of 10   Answer: Adding 40 will bring us
     will bring us to 85 or just   to 76.
     below?
B.   76 plus what equals 85?   Answer: 9
C.   What is 40 + 9?   Answer: 49 (the solution)

Math-Master Application

You are enjoying a 115-mile trip to the mountains, and your trip odometer currently reads 78 miles. How many miles are you away from your destination?

Calculation: 115 − 78 = ?

To subtract 78 from 115, answer the following questions

A.   78 plus what multiple of 10   Answer: Adding 30 will bring us
     will bring us to 115 or just   to 108.
     below?
B.   108 plus what equals 115?   Answer: 7
C.   What is 30 + 7?   Answer: 37 (the solution)

Food for Thought

Because this technique has been explained slowly, in a step-by-step fashion, it may seem to be rather time-consuming. However, with practice, you'll be able to process all the steps in your head simultaneously, producing an answer in only a second or two. Technique #6, coming up next, consists of a variation on Technique #5.

NOW IT'S YOUR TURN

Basic Exercises

1. 73
   − 28

2. 58
   − 36

3. 84
   − 47

4. 61
   − 43

5. 80
   − 43

6. 69
   − 21

7. 46
   − 17

8. 91
   − 55

9. 88
   − 35

10. 53
    − 29

11. 64
    − 38

12. 41
    − 17

Math Masters

13. 114
    − 77

14. 129
    − 54

15. 146
    − 87

16. 131
    − 66

17. 105
    − 57

18. 134
    − 88

19. 140
    − 73

20. 113
    − 69

(See answers on page 191)

DAY 2

TECHNIQUE 6:
Subtracting by Adding II

BASIC APPLICATION

Your team has just defeated the opposing team in a game of basketball by a score of 121–85. By how many points did your team win?

Calculation: $121 - 85 = ?$

The Old-Fashioned Way

$$\begin{array}{r} \not{1}\not{2}\not{1} \\ -\ 85 \\ \hline 36 \end{array}$$

The Excell-erated Way

This variation on Technique #5 is useful when the top number (minuend) is above 100 and the bottom number (subtrahend) is below 100. The key to this trick is determining the numerical distance that each number is from 100. We then add the two distances (amounts) to obtain the answer. Picture two numbers on a number line, one above 100 and the other below it, and you'll understand why this trick works. Let's go back to our basketball example and answer the question.

To subtract 85 from 121, answer the following questions

A.   121 is how much above 100?   Answer: 21
B.   85 is how much below 100?   Answer: 15
C.   What is 21 + 15?   Answer: 36 (the solution)

Let's Try One More Basic Calculation: 144 − 78

To subtract 78 from 144, answer the following questions

A.   144 is how much above 100?   Answer: 44
B.   78 is how much below 100?   Answer: 22
C.   What is 44 + 22?   Answer: 66 (the solution)

Math-Master Application

Though not mentioned above, this technique will also work with multiples of 100. Let's test it. Out of your freshman class of 425, 367 went on to graduate. How many (unfortunately) did not?

Calculation: 425 − 367 = ?

To subtract 367 from 425, answer the following questions

A.   425 is how much above 400?   Answer: 25
B.   367 is how much below 400?   Answer: 33
C.   What is 25 + 33?   Answer: 58 (the solution)

Food for Thought

If the Math-Master calculation above had been 425 − 267, we would simply take the distance (amount) above and below either 300 or 400 (take your pick), and add these amounts to obtain the answer. With practice (and use of Technique #5), you'll be able to perform Step B (determining the distance below a multiple of 100) in a flash.

NOW IT'S YOUR TURN

Basic Exercises

1.	111 − 93	2.	103 − 54	3.	116 − 88	4.	130 − 58
5.	151 − 86	6.	105 − 49	7.	165 − 78	8.	143 − 89
9.	137 − 93	10.	121 − 84	11.	114 − 77	12.	134 − 66

Math Masters

13.	530 − 477	14.	313 − 287	15.	631 − 575	16.	802 − 775
17.	428 − 395	18.	706 − 658	19.	939 − 880	20.	1,006 − 969

(See answers on page 192)

NUMBER MADNESS

Here's a strange one. Take any number, any number at all, and subtract the sum of the number's digits. What you'll get, without fail, is an answer that is evenly divisible by 9. For example, take the number 83,154. The sum of the digits $(8 + 3 + 1 + 5 + 4)$ is 21. Subtract 21 from 83,154, and you get 83,133, which is evenly divisible by 9!

DAY 2

TECHNIQUE 7:
Subtracting by Altering

BASIC APPLICATION

Your daughter is saving up to purchase $72 worth of computer software. She now has exactly $49. How much more must she save before she can make the purchase?

Calculation: $72 − $49 = ?

The Old-Fashioned Way

$$\begin{array}{r} \overset{6\ \ 12}{\cancel{7}\cancel{2}} \\ -\ 49 \\ \hline 23 \end{array}$$

The Excell-erated Way

You may recall that Technique #4 had you adding by altering. Now you are going to subtract in a similar fashion, but you're going to alter both the top number (minuend) and the bottom number (subtrahend). In our software example, 49 is a difficult number to subtract, but 50 (a multiple of 10) is much easier. Accordingly, we add 1 to both the 72 and the 49, producing the easier calculation, 73 − 50. Let's take a closer look at this nifty technique.

29

To subtract 49 from 72, answer the following questions

A. How far is 49 from the closest   Answer: 1 (below 50)
    multiple of 10?
B. What is 72 + 1?   Answer: 73
C. What is 49 + 1?   Answer: 50
D. What is 73 − 50?   Answer: 23 (the solution)

SUMMARY: 72 − 49 = (72 + 1)−(49 + 1) = 73 − 50 = 23

Let's Try One More Basic Calculation: 65 − 28

To subtract 28 from 65, answer the following questions

A. How far is 28 from the closest   Answer: 2 (below 30)
    multiple of 10?
B. What is 65 + 2?   Answer: 67
C. What is 28 + 2?   Answer: 30
D. What is 67 − 30?   Answer: 37 (the solution)

SUMMARY: 65 − 28 = (65 + 2) − (28 + 2) = 67 − 30 = 37

Math-Master Application

This technique will also work if the subtrahend is just *above* a multiple of 10. Let's assume you just read in one of the tabloids that an 81-year-old celebrity has a new 32-year-old spouse. Exactly what is the age difference?

Calculation: 81 − 32 = ?

To subtract 32 from 81, answer the following questions

A. How far is 32 from the closest   Answer: 2 (above 30)
    multiple of 10?
B. What is 81 − 2?   Answer: 79
C. What is 32 − 2?   Answer: 30
D. What is 79 − 30?   Answer: 49 (the solution)

SUMMARY: 81 − 32 = (81 − 2) − (32 − 2) = 79 − 30 = 49

Food for Thought

This technique is best used when the subtrahend is 1 or 2 *above* or *below* a multiple of 10. The key is to convert the subtrahend to the nearest multiple of 10 to make the calculation easier to perform. Remember that it is okay to alter the subtrahend as long as you *alter the minuend by the same amount in the same direction.*

NOW IT'S YOUR TURN

Basic Exercises

1.	54 − 19	2.	95 − 58	3.	100 − 69	4.	71 − 38
5.	67 − 39	6.	74 − 38	7.	46 − 29	8.	52 − 18
9.	73 − 38	10.	86 − 49	11.	64 − 18	12.	94 − 59

Math Masters

13.	114 − 79	14.	90 − 19	15.	83 − 28	16.	106 − 68
17.	91 − 29	18.	112 − 89	19.	105 − 78	20.	87 − 18

(See answers on page 192)

DAY 2

TECHNIQUE 8:
Subtracting in Two Steps

BASIC APPLICATION

After waiting for a sale at the local electronics store, you buy a telephone answering machine for $56 that normally sells for $90. How much did you save by waiting for the sale?

Calculation: $90 − $56 = ?

The Old-Fashioned Way

$$\begin{array}{r} \overset{8\ \ 10}{\cancel{9}\cancel{0}} \\ -\ 56 \\ \hline 34 \end{array}$$

The Excell-erated Way

This technique is similar to Techniques #2 and #3, in that you must pay careful attention to each digit's place value. The strategy is to first subtract the tens digit and then subtract the ones digit. Using our answering-machine example, we would first subtract 50 from 90, and then subtract the 6. In other words, 56 is to be viewed as (50 + 6). Let's take this example one step at a time.

To subtract 56 from 90, answer the following questions

A.   What is another way to view the   Answer: $(50 + 6)$
subtrahend?

B.   What is $90 - 50$?   Answer: 40

C.   What is $40 - 6$?   Answer: 34 (the solution)

SUMMARY: $90 - 56 = 90 - (50 + 6) = 90 - 50 - 6 = 34$

Let's Try One More Basic Calculation: $110 - 37$

To subtract 37 from 110, answer the following questions

A.   What is another way to view the   Answer: $(30 + 7)$
subtrahend?

B.   What is $110 - 30$?   Answer: 80

C.   What is $80 - 7$?   Answer: 73 (the solution)

SUMMARY: $110 - 37 = 110 - (30 + 7) = 110 - 30 - 7 = 73$

Math-Master Application

You've just purchased something at a garage sale for $83, and made a $25 deposit to hold the item. When you return with the balance of the payment, how much will you owe?

Calculation: $\$83 - \$25 = ?$

To subtract $25 from $83, answer the following questions

A.   What is another way to view the   Answer: $(20 + 5)$
subtrahend?

B.   What is $83 - 20$?   Answer: 63

C.   What is $63 - 5$?   Answer: 58 (the solution)

SUMMARY: $83 - 25 = 83 - (20 + 5) = 83 - 20 - 5 = 58$

Food for Thought

A variation on this technique is to *first* subtract the ones digit and *then* the tens. Using our Math-Master example, we could first take $83 - 5$, which equals 78. Then we could take $78 - 20$, which equals our answer,

58. It's probably easier, however, to start with the tens digit, and then subtract the ones digit. This technique will also work with larger numbers, such as those in the hundreds. However, you need a lot of practice and even more concentration to pull it off.

NOW IT'S YOUR TURN

Basic Exercises

1. $80 - 46 =$
2. $120 - 59 =$
3. $90 - 34 =$
4. $100 - 23 =$
5. $60 - 21 =$
6. $70 - 47 =$
7. $130 - 62 =$
8. $80 - 29 =$
9. $50 - 18 =$
10. $140 - 83 =$
11. $90 - 24 =$
12. $110 - 61 =$
13. $200 - 55 =$

Math Masters

14. $61 - 27 =$
15. $93 - 54 =$
16. $84 - 18 =$
17. $97 - 39 =$
18. $131 - 46 =$

19. $165 - 77 =$

20. $203 - 155 =$

(See answers on page 192)

JUST FOR FUN

How would you go about subtracting the squares of two consecutive whole numbers? For example, what does $9^2 - 8^2$ or $25^2 - 24^2$ equal? The trick—simply add the consecutive numbers and you've got the answer. So $9^2 - 8^2 = 9 + 8$, which equals 17. Also, $25^2 - 24^2 = 25 + 24$, which equals 49. Seems almost too good to be true!

DAY 2 REVIEW QUIZ

Directions: Work each exercise as quickly as possible, using the techniques just learned.

Basic Exercises

1.	63 − 29	**2.**	114 − 87	**3.**	52 − 18	**4.**	90 − 43
5.	131 − 93	**6.**	86 − 39	**7.**	120 − 71	**8.**	80 − 24

Math Masters

9.	134 − 89	**10.**	531 − 475	**11.**	106 − 78	**12.**	142 − 86

(See answers on page 192)

TOTALING TIP

When adding a small handful of numbers, it's usually easiest to begin with the largest number, and work your way down. For example, see how difficult it is to add $7 + 15 + 29 + 56$. However, taking $56 + 29 + 15 + 7$ makes the calculation much simpler. Try it on some other numbers, and you'll see!

DAY 3:

MULTIPLICATION AND DIVISION I

DAY 3

TECHNIQUE 9:
Multiplying by 4

BASIC APPLICATION

Chess Cab Company owns and operates 23 taxis. The owner discovers a really good deal on tires, and decides to replace all four tires for each of the 23 vehicles. How many tires will the owner be purchasing?

Calculation: $23 \times 4 = ?$

The Old-Fashioned Way

$$\begin{array}{r} \overset{1}{2}3 \\ \times\,4 \\ \hline 92 \end{array}$$

Note the multiplication in two steps, from right to left, requiring some carrying.

The Excell-erated Way

Multiplying a number by 4 is identical with doubling the number and then doubling again. There is no need to carry, and you can perform the calculation quickly and easily in your head. Using the example above, $23 \times 4 = (23 \times 2) \times 2$. Read on for a step-by-step explanation.

To multiply 23 by 4, answer the following questions

A. What is 23×2?   Answer: 46
B. What is 46×2?   Answer: 92 (the solution to 23×4)

Let's Try One More Basic Calculation: 57 × 4

To multiply 57 by 4, answer the following questions

A. What is 57 × 2?   Answer: 114
B. What is 114 × 2?   Answer: 228 (the solution to 57 × 4)

Math-Master Application

This technique will also work when multiplying a number by 0.4, 40, 400, and so on. Let's assume you've made a purchase that must be satisfied with 40 British pounds. You only have American dollars, and must produce $1.80 on that day to obtain one British pound. How many American dollars will you need to acquire the 40 British pounds needed for payment?

Calculation: $1.80 × 40 = ?

To multiply $1.80 by 40, answer the following questions

A. What does the calculation look   Answer: 18 × 4
    like, omitting the dollar sign,
    decimal point, and two
    zeroes?
B. What is 18 × 2?   Answer: 36
C. What is 36 × 2?   Answer: $72 (tentative solution,
                                dollar sign added)
D. Let's now test the tentative solution, $72, to see if it appears to be the solution to $1.80 × 40. We know that $1.80 is close to $2, and $2 × 40 = $80. So the solution to $1.80 × 40 must be just under $80, or $72. Another way to test the tentative solution is to alter it by inserting a decimal point or tacking on a zero or two. For example, $7.20 would "seem" too small to be the solution, while $720 is definitely too large. Therefore, you can determine by inspection that the tentative solution of $72 is in fact the answer to $1.80 × 40.

Food for Thought

The smaller the number, the more likely you will encounter it in a calculation. For example, the number 7 will enter into more calculations than the number 846. Therefore, the above technique for multiplying by 4 will come in handy time and time again because the number 4 is so small. To master this trick, practice doubling every number in sight. For example, when you see a two- or three-digit number on a billboard or a license plate, double it in your head. With practice, it will become second nature to you!

NOW IT'S YOUR TURN

Basic Exercises

1. $16 \times 4 =$
2. $27 \times 4 =$
3. $73 \times 4 =$
4. $4 \times 52 =$
5. $4 \times 38 =$
6. $4 \times 64 =$
7. $88 \times 4 =$
8. $19 \times 4 =$
9. $93 \times 4 =$
10. $4 \times 31 =$
11. $4 \times 45 =$
12. $4 \times 66 =$
13. $84 \times 4 =$

Math Masters

14. $28 \times 40 =$
15. $460 \times 0.4 =$
16. $9.5 \times 4 =$
17. $0.35 \times 400 =$
18. $0.4 \times 710 =$

19. $40 \times 5.9 =$

20. $99 \times 4 =$

(See answers on page 192)

DAY 3

TECHNIQUE 10:
Dividing by 4

BASIC APPLICATION

You've just purchased four reams of paper at the local office supply warehouse, and paid a total of $72. How much did each ream cost?

Calculation: $72 ÷ 4 = ?

The Old-Fashioned Way

$$\begin{array}{r} 18 \\ 4\overline{)72} \\ \underline{4} \\ 32 \\ \underline{32} \\ 0 \end{array}$$

Cumbersome!
Cumbersome!
Cumbersome!

The Excell-erated Way

Dividing a number by 4 is identical with cutting the number in half, and then cutting it in half again. Even though two steps are needed, with practice you can perform each step quickly and easily in your head. Using the example above, $72 ÷ 4 = $72 ÷ 2 ÷ 2. Read on for a step-by-step explanation.

To divide 72 by 4, answer the following questions

A. What is 72 ÷ 2?   Answer: 36
B. What is 36 ÷ 2?   Answer: $18 (the solution to $72 ÷ 2)

Let's Try One More Basic Calculation: 96 ÷ 4

To divide 96 by 4, answer the following questions

A.   What is 96 ÷ 2?   Answer: 48
B.   What is 48 ÷ 2?   Answer: 24 (the solution to 96 ÷ 4)

Math-Master Application

This technique will also work when dividing a number by 0.4, 40, 400, and so on. To illustrate, you've just purchased 400 shares of stock for $8,200. What is the cost per share?

Calculation: $8,200 ÷ 400 = ?

To divide $8,200 by 400, answer the following questions

A.   What does the calculation   Answer: 82 ÷ 4
     look like, omitting the dollar
     sign and zeroes?
B.   What is 82 ÷ 2?   Answer: 41
C.   What is 41 ÷ 2?   Answer: 20.5, or $20.50
                                       (tentative solution)
D.   Let's now test the tentative solution, $20.50, to see if it appears to be
     the solution to $8,200 ÷ 400. Remember that when dividing, you
     can eliminate an equal number of zeroes from the dividend and
     divisor. Thus, you have $82 ÷ 4, to which the tentative solution of
     $20.50 is clearly the solution. Another way to test the tentative
     solution is to move the decimal point left and right, producing a too
     low possibility of $2.05 and a too high possibility of $205.

Food for Thought

Because this technique involves such a small number (4), it will come
into play quite often. As was recommended with Technique #9, practice
as much as you can, but this time cutting every number in sight in half.
For example, cut the numbers of a basketball score, or the prefix of your
telephone number, or perhaps your street number, in half for practice.
When you learn Techniques #15 and #16 (Day 4), you'll realize how
important Techniques #9 and #10 really are!

NOW IT'S YOUR TURN

Basic Exercises

1. $92 \div 4 =$
2. $56 \div 4 =$
3. $116 \div 4 =$
4. $220 \div 4 =$
5. $84 \div 4 =$
6. $76 \div 4 =$
7. $132 \div 4 =$
8. $480 \div 4 =$
9. $64 \div 4 =$
10. $260 \div 4 =$
11. $128 \div 4 =$
12. $180 \div 4 =$
13. $68 \div 4 =$

Math Masters

14. $62 \div 4 =$
15. $140 \div 40 =$
16. $3,400 \div 400 =$
17. $86 \div 4 =$
18. $10.4 \div 0.4 =$
19. $1,440 \div 40 =$
20. $18.4 \div 4 =$

(See answers on page 192)

QUICK QUESTION

Can you figure out a technique to quickly multiply any two-digit number by 101? For example, how could you instantly determine the answer to 74 × 101?

ANSWER: Simply write the number being multiplied by 101 *twice*. For example, 74 × 101 = 7,474.

DAY 3

TECHNIQUE 11:
Multiplying by 5

BASIC APPLICATION

You have just joined a bowling league, which consists of 28 teams of five bowlers each. How many bowlers are there in the league?

Calculation: $28 \times 5 = ?$

The Old-Fashioned Way

$$\begin{array}{r} \overset{4}{28} \\ \times 5 \\ \hline 140 \end{array}$$

The Excell-erated Way

Multiplying a number by 5 is identical with cutting the number in half, and then multiplying by 10. Using the example above, $28 \times 5 = (28 \div 2) \times 10$. Remember that to multiply a number by 10, simply tack on a zero to a whole number, or if the number has a decimal point, move the decimal point one place to the right.

To multiply 28 by 5, answer the following questions

A. What is half of 28?   Answer: 14 (a whole number)
B. What is 14×10?   Answer: 140 (the solution to 28×5)

Let's Try One More Basic Calculation: 33×5

To multiply 33 by 5, answer the following questions

A.   What is half of 33?   Answer: 16.5.
B.   What is 16.5×10?   Answer: 165 (the solution to 33×5)

Remember: To multiply a number with a decimal point by 10, simply move the decimal point one place to the right.

Math-Master Application

This technique will also work when multiplying by 0.5, 50, 500, and so on. Let's assume that you have been employed by your company long enough to earn paid vacation time. Specifically, you earn $73 of vacation pay each week for the 50 weeks you work per year. How much vacation pay do you accrue per year?

Calculation: $\$73 \times 50 = ?$

To multiply $73 by 50, answer the following questions

A.   What does the calculation   Answer: 73×5
     look like, omitting the dollar
     sign and zero?
B.   What is half of 73?   Answer: 36.5, or $36.50 (tentative
                                   solution)

Obviously, $36.50 is much too small to be the solution to $\$73 \times 50$. Moving the decimal point one place to the right produces $365, which would be the answer to $\$73 \times 5$. However, moving the decimal *two* places to the right produces the correct answer, $3,650.

Food for Thought

An alternative way to multiply by 5 is to reverse the steps explained above. That is, first multiply the number by 10, and *then* cut in half. Using the Math-Master example above, 73×5 could be computed by first taking 73×10, or 730, and then cutting 730 in half, producing the answer (or tentative answer) 365. Use whichever application you find easier. In the end, remember that multiplication by 5 is the same as multiplication by 10, and then division by 2.

NOW IT'S YOUR TURN

Basic Exercises

1. $36 \times 5 =$
2. $64 \times 5 =$
3. $46 \times 5 =$
4. $5 \times 24 =$
5. $5 \times 78 =$
6. $5 \times 52 =$
7. $84 \times 5 =$
8. $72 \times 5 =$
9. $98 \times 5 =$
10. $5 \times 56 =$
11. $5 \times 88 =$
12. $5 \times 42 =$
13. $68 \times 5 =$

Math Masters

14. $45 \times 5 =$
15. $77 \times 5 =$
16. $5 \times 39 =$
17. $50 \times 5.8 =$
18. $0.5 \times 920 =$
19. $6.6 \times 500 =$
20. $260 \times 50 =$

(See answers on page 193)

DAY 3

TECHNIQUE 12:
Dividing by 5

BASIC APPLICATION

You're at a restaurant with four friends, and you have just finished eating dessert. The check, including tax and tip, totals $64. The five of you agree to divide the check equally, and that you will be the one to perform the calculation. How much will each of you pay?

Calculation: $64 ÷ 5 = ?

The Old-Fashioned Way

$$
\begin{array}{r}
12.80 \\
5)\overline{64.00} \\
5 \\
\overline{14} \\
10 \\
\overline{40} \\
40 \\
\overline{0}
\end{array}
$$

The Excell-erated Way

Dividing a number by 5 is the same as doubling the number, and then dividing by 10. Using the example above, $64 \div 5 = (64 \times 2) \div 10$. Remember that to divide a number by 10, simply move the decimal point one place to the left. The theory behind this technique is that it is easier to multiply by 2 than to divide by 5. Let's look below to see how this technique works.

50

To divide $64 by 5, answer the following questions

A. What is 64 × 2? Answer: 128
B. What is 128 ÷ 10? Answer: 12.8, or $12.80 (the solution)

Let's Try One More Basic Calculation: 29 ÷ 5

To divide 29 by 5, answer the following questions

A. What is 29 × 2? Answer: 58
B. What is 58 ÷ 10? Answer: 5.8 (the solution to 29 ÷ 5)

Math-Master Application

This technique will also work when dividing by 0.5, 50, 500, and so on. Let's assume that there are 9,200 members of your professional organization in the United States, and you'd like to know the average number of members per state. What is the answer?

Calculation: 9,200 ÷ 50 = ?

To divide 9,200 by 50, answer the following questions

A. What does the calculation Answer: 92 ÷ 5
   look like, omitting the zeroes?
B. What is 92 × 2? Answer: 184 (tentative solution)

Let's now test the tentative solution, 184, to see if it appears to be the solution to 9,200 ÷ 50. First, let's eliminate one zero each from the dividend and divisor, producing 920 ÷ 5. Next, remember that division is the inverse of multiplication. Thus, working backwards, ask yourself, "Does 184 times 5 seem to equal 920?" The answer is yes, so the tentative solution, 184, is also the final solution.

Food for Thought

An alternative way to divide by 5 is to reverse the steps explained above. That is, first divide the number by 10, and *then* multiply by 2. Using the example that introduced this technique, 64 ÷ 5 could be computed by first taking 64 ÷ 10, or 6.4, and then doubling, producing the answer 12.8 (or $12.80). Use whichever application you find easier. In the end, remember that division by 5 is the same as division by 10, and then multiplication by 2.

NOW IT'S YOUR TURN

Basic Exercises

1. $43 \div 5 =$
2. $71 \div 5 =$
3. $56 \div 5 =$
4. $22 \div 5 =$
5. $87 \div 5 =$
6. $36 \div 5 =$
7. $94 \div 5 =$
8. $103 \div 5 =$
9. $48 \div 5 =$
10. $17 \div 5 =$
11. $76 \div 5 =$
12. $112 \div 5 =$
13. $31 \div 5 =$

Math Masters

14. $340 \div 50 =$
15. $28.5 \div 5 =$
16. $7,300 \div 500 =$
17. $22.2 \div 5 =$
18. $13.5 \div 0.5 =$
19. $390 \div 50 =$
20. $810 \div 500 =$

(See answers on page 193)

JUST FOR FUN

One of the most difficult tasks in writing this book was to determine the best order for the 40 Excell-erated techniques. How would you calculate the number of different ways the 40 techniques could have been arranged?

ANSWER: Determining the number of different ways a group of items can be arranged, such as books on a bookshelf, involves the concept of factorial. For example, 6 factorial (written "6!") means $1 \times 2 \times 3 \times 4 \times 5 \times 6$, which equals 720. Therefore, the 40 techniques in this book can be arranged in 40! ways ($1 \times 2 \times \ldots \times 40$). This product turns out to be about 8.16×10^{47}, or about 816 quattuordecillion (a number with 48 digits).

DAY 3 REVIEW QUIZ

DIRECTIONS: Work each exercise as quickly as possible, using the techniques just learned.

Basic Exercises

1.	$45 \times 4 =$	5.	$78 \div 5 =$
2.	$260 \div 4 =$	6.	$5 \times 72 =$
3.	$42 \times 5 =$	7.	$128 \div 4 =$
4.	$36 \div 5 =$	8.	$4 \times 27 =$

Math Masters

9. $3.5 \times 40 =$

11. $8.8 \times 50 =$

10. $3,800 \div 400 =$

12. $1,120 \div 500 =$

(See answers on page 193)

QUICK CHECK

Want to know how to quickly check multiplication by 9, 99, 999, and so forth? Well, there's an amazingly easy way. All you have to do is add up the digits of the product. If that sum is evenly divisible by 9, your answer is *probably correct*. If not, then your answer is *definitely incorrect*. For example, the calculation "99 × 230 = 22,870" is definitely incorrect because the digit-sum of the product (19) is not evenly divisible by 9. However, the calculation "78 × 99 = 77,922" is probably correct because the digit-sum of the product (27) is evenly divisible by 9. The reason it is probably, but not definitely, correct is that it is possible to produce an incorrect answer whose digit-sum *is* evenly divisible by 9. In the "78 × 99" example above, an answer of 77,292 is incorrect, but still produces a digit-sum of 27.

DAY 4:

MULTIPLICATION AND DIVISION II

DAY 4

TECHNIQUE 13:
Multiplying Any Two-Digit Number by 11

BASIC APPLICATION

You are considering the purchase of an economy car which averages 53 miles per gallon and contains an 11-gallon gas tank. Assuming you begin with a full tank, how many miles would you expect to be able to drive before running out of gas?

Calculation: $53 \times 11 = ?$

The Old-Fashioned Way

```
   53
 × 11
   53
   53
  583
```

The Excell-erated Way

This trick is everyone's favorite because it's so easy to execute. To multiply a number by 11, simply separate the two digits of the number and place the sum of the digits in the middle. Using the example above, the digits of the number 53 total 8 (i.e., 5 + 3). Place the 8 between the 5 and the 3 and you've got the answer to 53×11. If the sum of the digits is more than 9, as in the number 86, simply carry to the left digit. Let's try a few.

To multiply 53 by 11, answer the following questions

A.   What is 5 + 3?   Answer: 8
B.   What is 53 with the 8   Answer: 583 (the solution to
     inserted in the middle?   53 × 11)

Let's Try One More Basic Calculation: 86 × 11

To multiply 86 by 11, answer the following questions

A.   What is 8 + 6?   Answer: 14 (insert the 4, carry
                                                          the 1)

B.   What is 86 with the 4   Answer: 846
     inserted in the middle?

C.   What is 846 with the 1   Answer: 946 (the solution to
     carried to the hundreds   86 × 11)
     digit?

Math-Master Application

This technique will also work when multiplying by 0.11, 1.1, 110, and so on. Let's assume your monthly telephone bill averages $43. Because of an approved increase in rates, you expect your phone bills to rise by 10 percent. What, therefore, do you expect your upcoming bills will average?

Calculation: $43 × 1.1 = ?
                    ($43 × 110%)

To multiply $43 by 1.1, answer the following questions

A.   What does the calculation   Answer: 43 × 11
     look like, omitting the dollar
     sign and decimal point?
B.   What is 4 + 3?   Answer: 7
C.   What is 43 with the 7   Answer: 473 (tentative solution)
     inserted in the middle?

Now let's test the tentative solution, 473, to see if it appears to be the solution to $43 × 1.1. We know that 1.1 is slightly greater than 1, so $43 × 1.1 must equal slightly more than $43. If we insert a decimal point between the 7 and the 3 of the tentative solution, and attach a dollar sign and right-hand zero, we produce the final solution, $47.30.

Food for Thought

There is no right or wrong way to test a tentative solution to see if a decimal point should be inserted or zeroes tacked on. The methods explained in this book are merely suggestions, and you should consider performing a different test if it works better for you. The key is to see if your solution makes sense, given the original calculation. For example, in the Math Master computation above, you should be able to see plainly that $473 would be much too large to be the solution to $43 × 1.1, whereas $4.73 would be impossibly small.

NOW IT'S YOUR TURN

Basic Exercises

1. $25 \times 11 =$
2. $43 \times 11 =$
3. $81 \times 11 =$
4. $11 \times 33 =$
5. $11 \times 62 =$
6. $11 \times 59 =$
7. $28 \times 11 =$
8. $75 \times 11 =$
9. $96 \times 11 =$
10. $11 \times 36 =$
11. $11 \times 16 =$
12. $11 \times 67 =$
13. $98 \times 11 =$

Math Masters

14. $2.3 \times 110 =$

15. $93 \times 1.1 =$

16. $1.1 \times 4.8 =$

17. $1,100 \times 0.79 =$

18. $110 \times 410 =$

19. $550 \times 110 =$

20. $840 \times 0.11 =$

(See answers on page 193)

DAY 4

TECHNIQUE 14:
Squaring Any Number Ending in 5

BASIC APPLICATION

Because your family room experiences a lot of traffic, the carpeting is badly in need of replacement. You've decided to replace the carpeting with one-foot-square tiles. If your family room measures exactly 25' by 25', how many tiles will you need to purchase to cover the entire room?

Calculation: $25 \times 25 = ?$

The Old-Fashioned Way

$$
\begin{array}{r}
1\ 2\\
25\\
\times\ 25\\
\hline
125\\
50\\
\hline
625
\end{array}
$$

The Excell-erated Way

This trick is another winner because it's so much fun to apply. When a number ending in 5 is squared (multiplied by itself), the solution automatically ends in 25. To obtain the left-hand portion of the solution, multiply the tens digit of the number by the next whole number. In the example above, the tens digit is 2, and the next whole number is 3, so the product of 2×3 is 6. As you're about to see, it's a lot easier than it sounds.

To multiply 25 by 25, answer the following questions

A. What does this type of   Answer: 25
    calculation automatically end
    in?

B. What is the tens digit   Answer: $2 \times 3 = 6$
    multiplied by the next whole
    number?

C. What do you get when you   Answer: 625 (the solution to
    combine the 6 and the 25? 25×25)

Let's Try One More Basic Calculation: 75×75

To multiply 75 by 75, answer the following questions

A. What does this type of   Answer: 25
    calculation automatically end
    in?

B. What is the tens digit   Answer: $7 \times 8 = 56$
    multiplied by the next whole
    number?

C. What do you get when you   Answer: 5,625 (the solution to
    combine the 56 and the 25? 75×75)

Math-Master Application

Even though a calculation such as 450×4.5 is not technically a square, this technique will still work. Let's say you've decided to have the outside of your house repainted. Total estimated labor time is 95 hours, and your house painter charges $9.50 per hour. What is the expected total labor cost of having your house painted?

Calculation: $9.50 $\times$ 95 = ?

To multiply $9.50 by 95, answer the following questions

A. What does the calculation   Answer: 95×95
    look like, omitting the dollar
    sign, decimal point, and
    zero?

B. What does this sort of   Answer: 25
    calculation automatically end
    in?

C. What is the tens digit   Answer: $9 \times 10 = 90$
    multiplied by the next whole
    number?

D. What do you get when you   Answer: 9,025, or $9,025
    combine the 90 and the 25?   (tentative solution)

Now let's test the tentative solution, $9,025, to see if it appears to be the solution to $9.50 × 95. We know that $9.50 is just shy of $10, and that $10 × 95 = 950. So the solution to $9.50 × 95 must be a bit less than $950. If we move the decimal of the tentative solution one place to the left, we obtain the final solution, $902.50. We could have tested the tentative solution by instead rounding up the labor hours to 100, and proceeding in the same manner.

Food for Thought

Why not combine Techniques #13 and #14 by squaring the number 115? You know that the answer automatically ends in 25. Then, take 11 times the next whole number (12) to obtain 132. Combine the 132 and the 25 to obtain the answer, 13,225! As we continue with this 10-day program, don't hesitate to use previously learned techniques when appropriate. Remember, the objective is to obtain the answer as quickly as possible (without a calculator, of course). How you choose to proceed is entirely up to you.

NOW IT'S YOUR TURN

Basic Exercises

1. $65^2 =$

2. $35^2 =$

3. $85^2 =$

4. $15 \times 15 =$

5. $95 \times 95 =$

6. $45 \times 45 =$

7. $75^2 =$

8. $55^2 =$

9. $25^2 =$

10. $105 \times 105 =$

11. $35 \times 35 =$

12. $65 \times 65 =$

13. $115^2 =$

Math Masters

14. $150 \times 1.5 =$

15. $350^2 =$

16. $8.5 \times 8.5 =$

17. $5.5 \times 550 =$

18. $0.95 \times 0.95 =$

19. $450 \times 45 =$

20. $650^2 =$

(See answers on page 193)

QUICK CHALLENGE

Now that you know the secret to mentally multiplying by 11, let's see if you can figure out how to quickly multiply by 111. For example, we know instantly that $35 \times 11 = 385$. But what about 35×111?

ANSWER: Use the same trick as multiplying 35 by 11, except insert another "8" in the middle, producing the answer 3,885.

DAY 4

TECHNIQUE 15:
Multiplying by 25

BASIC APPLICATION

You can never find a pencil in the house, so you go down to the local stationery store and buy 36 boxfuls. If each box contains 25 pencils, how many have you purchased in all?

Calculation: $36 \times 25 = ?$

The Old-Fashioned Way

$$
\begin{array}{r}
1\ 3 \\
36 \\
\times\ 25 \\
\hline
180 \\
72 \\
\hline
900
\end{array}
$$

The Excell-erated Way

This technique is similar to Technique #11 (Multiplying by 5) in that you are going to multiply by dividing. Specifically, multiplying a number by 25 is the same as dividing the number by 4, and then multiplying by 100 (tacking on two zeroes or moving the decimal point to the right two places). The theory here is that it is easier to divide by 4 than it is to multiply by 25, especially now that you've learned Technique #10 (Dividing by 4). Let's see how this technique applies to our pencil problem stated above.

To multiply 36 by 25, answer the following questions

A.   What is 36 ÷ 4?   Answer: 9
B.   What is 9 × 100?   Answer: 900 (the solution to
                                  36 × 25)

Let's Try One More Basic Calculation: 52 × 25

To multiply 52 by 25, answer the following questions

A.   What is 52 ÷ 4?   Answer: 13 (Did you use
                                  Technique #10 here?)

B.   What is 13 × 100?   Answer: 1,300 (the solution to
                                  52 × 25)

Math-Master Application

This technique will also work when multiplying by 0.25, 2.5, 250, and so on. Also note that 2.5 can also be expressed as 2½. Let's assume that a textbook publisher ships 44 math books to a college bookstore. If each book weighs 2½ lbs., what is the total weight of the shipment?

Calculation: 44 × 2.5 = ?

To multiply 44 by 2.5, answer the following questions

A.   What does the calculation   Answer: 44 × 25
     look like, omitting the
     decimal point?
B.   What is 44 ÷ 4?   Answer: 11 (tentative solution)

Now let's test the tentative solution, 11, to see if it appears to be the solution to 44 × 2.5. Simply by inspection, we can determine that the solution must be slightly over 100. If we then tack on a zero to the tentative solution, we obtain the final solution, 110.

Food for Thought

As you may have already guessed, the technique can be applied in reverse. That is, to multiply a number by 25, you can first multiply the number by 100, and *then* divide by 4. Using the original pencil calculation (36 × 25), we could multiply 36 by 100, to produce 3,600. Then we could divide 3,600 by 4 to obtain the final solution 900. For most people, however, it is easier first to divide by 4 and then multiply by 100.

NOW IT'S YOUR TURN

Basic Exercises

1. 76 × 25 =
2. 44 × 25 =
3. 32 × 25 =
4. 25 × 16 =
5. 25 × 68 =
6. 25 × 88 =
7. 92 × 25 =
8. 28 × 25 =
9. 56 × 25 =
10. 25 × 24 =
11. 25 × 64 =
12. 25 × 48 =
13. 96 × 25 =

Math Masters

14. 46 × 25 =
15. 54 × 25 =
16. 25 × 62 =
17. 720 × 2.5 =
18. 250 × 14 =
19. 86 × 0.25 =
20. 2.5 × 84 =

(See answers on page 193)

DAY 4

TECHNIQUE 16:
Dividing by 25

BASIC APPLICATION

You've just purchased a bicycle, and your goal is to ride it 220 miles per week. If you average 25 miles per hour on your bicycle, how many hours per week must you ride to reach your goal?

Calculation: $220 \div 25 = ?$

The Old-Fashioned Way

```
      8.8
25)220.0
   200
    20 0
    20 0
       0
```

The Excell-erated Way

Dividing a number by 25 is the same as multiplying the number by 4, and then dividing by 100 (which, you'll remember, is accomplished by moving the decimal point two places to the left). The theory is that it is easier to multiply by 4 than to divide by 25, especially now that you've learned Technique #9 (Multiplying by 4). Let's try this technique out on our bicycle example.

To divide 220 by 25, answer the following questions

A. What is 220 × 4?   Answer: 880 (better yet, take
                               22 × 4, and tack on a zero)

B. What is 880 ÷ 100?   Answer: 8.8 (the solution to
                               220 ÷ 25)

Let's Try One More Basic Calculation: 700 ÷ 25

To divide 700 by 25, answer the following questions

A. What is 700 × 4?   Answer: 2,800 (better yet, take
                               7 × 4, and tack on two zeroes)

B. What is 2,800 ÷ 100?   Answer: 28 (the solution to
                               700 ÷ 25)

Math-Master Application

You have just enrolled in a speed reading course. To test your current speed, your instructor has you read a 900-word passage, which takes you exactly 2½ (or 2.5) minutes to read. What is your reading speed upon beginning the course?

Calculation: 900 ÷ 2.5 = ?

To divide 900 by 2.5, answer the following questions

A. What does the calculation   Answer: 9 ÷ 25
    look like, omitting the
    decimal point and zeroes?
B. What is 9 × 4?   Answer: 36 (tentative solution)

Let's test the tentative solution, 36, to see if it appears to be the solution to 900 ÷ 2.5. Remembering that multiplication is the inverse of division, ask yourself, "Does 36 × 2.5 = 900?" Obviously, it does not. Try tacking a zero onto the 36, and then ask yourself, "Does 360 × 2.5 = 900?" No doubt about it, 360 is the final solution to 900 ÷ 2.5.

Food for Thought

If you prefer, you can apply Technique #16 in reverse. That is, to divide a number by 25, you can first divide it by 100, and *then* multiply by 4. Using the second example presented above, 700 ÷ 25, first divide 700 by 100, which equals 7. Then multiply 7 by 4, to produce the answer 28. Whichever alternative you choose, you can simply ignore any zeroes, multiply by 4, and then ask yourself where it would make sense to insert a decimal point or tack on a zero or two to produce the answer.

NOW IT'S YOUR TURN

Basic Exercises

1. $900 \div 25 =$
2. $110 \div 25 =$
3. $700 \div 25 =$
4. $3,200 \div 25 =$
5. $8,500 \div 25 =$
6. $410 \div 25 =$
7. $120 \div 25 =$
8. $550 \div 25 =$
9. $640 \div 25 =$
10. $300 \div 25 =$
11. $7,000 \div 25 =$
12. $1,700 \div 25 =$
13. $800 \div 25 =$

Math Masters

14. $450 \div 2.5 =$
15. $1,400 \div 250 =$
16. $2.1 \div 0.25 =$
17. $88 \div 25 =$
18. $47 \div 2.5 =$

19. $710 \div 250 =$

20. $19,000 \div 2,500 =$

(See answers on page 194)

HELPFUL HINT

Let's say you have to multiply 7 by 18 in your head. For most people, this is a fairly formidable task. However, you can divide the calculation into two steps: $7 \times 18 = 7 \times 9 \times 2$, which equals 63×2, which equals 126. This technique also works well when multiplying by 22. For example, $13 \times 22 = 13 \times 11 \times 2$. You know that $13 \times 11 = 143$ (Technique #13), and $143 \times 2 = 286$. Try this little trick whenever the multiplication seems a wee bit difficult.

DAY 4 REVIEW QUIZ

DIRECTIONS: Work each exercise as quickly as possible, using the techniques just learned.

Basic Exercises

1. $36 \times 11 =$ **5.** $700 \div 25 =$

2. $85 \times 85 =$ **6.** $25 \times 32 =$

3. $64 \times 25 =$ **7.** $45^2 =$

4. $120 \div 25 =$ **8.** $11 \times 85 =$

Math Masters

9. $1.1 \times 430 =$

10. $150 \times 1.5 =$

11. $2.5 \times 9.6 =$

12. $3,600 \div 250 =$

(See answers on page 194)

NUMBER CURIOSITY

$37,037,037 \times 3 \ = 111,111,111$
$37,037,037 \times 6 \ = 222,222,222$
$37,037,037 \times 9 \ = 333,333,333$
$37,037,037 \times 12 = 444,444,444$
$37,037,037 \times 15 = 555,555,555$
etc.

DAY 5:

MULTIPLICATION AND DIVISION III

DAY 5

TECHNIQUE 17:
Multiplying by 1.5, 2.5, 3.5, etc.

BASIC APPLICATION

Your local copy center charges 6½ (or 6.5) cents for each legal-size copy. You need to make 8 legal-size copies, but only have 55 cents on you and are afraid that that isn't enough to cover the cost. What will the 8 copies cost?

Calculation: $8 \times 6.5 = ?$

The Old-Fashioned Way

$$
\begin{array}{r}
4 \\
6.5 \\
\times\ \ 8 \\
\hline
52.0
\end{array}
$$

The Excell-erated Way

The assumption here is that it is easier to multiply by a whole number than by a mixed number (as 6½ or 6.5). The trick is to double the number ending in .5 (producing a whole number) and to cut the other number in half (hopefully producing another whole number). Using the example above, we would double the 6.5, producing 13, and cut the 8 in half, producing 4. Let's finish the calculation with a step-by-step explanation below.

To multiply 8 by 6.5, answer the following questions

A.   What is 6.5 × 2?   Answer: 13
B.   What is half of 8?   Answer: 4
C.   What is 13 × 4?   Answer: 52 (the solution to 6.5 × 8)

Let's Try One More Basic Calculation: 14 × 3.5

To multiply 14 by 3.5, answer the following questions

A.   What is 3.5 × 2?   Answer: 7
B.   What is half of 14?   Answer: 7
C.   What is 7 × 7?   Answer: 49 (the solution to 14 × 3.5)

Math-Master Application

You're about to take a long trip by car, and are reading a road map to estimate the total mileage. If each inch on your map represents 160 miles, and you've measured 4.5 inches between starting point and destination, approximately how many miles will you be driving?

Calculation: 160 × 4.5 = ?

To multiply 160 by 4.5, answer the following questions

A.   What is 4.5 × 2?   Answer: 9
B.   What is half of 160?   Answer: 80
C.   What is 9 × 80?   Answer: 720 (the solution to
                             160 × 4.5)

Food for Thought

Even though presentation of this technique has been limited to numbers ending in .5, it works with any number ending in 5. For example, if you wish to multiply 75 by 60, just double the 75 and halve the 60, producing the calculation 150 × 30 (which, of course, is the same as 15 × 3, with two zeroes tacked on, or 4,500). With practice, you will be able to apply this advanced application almost as easily as the basic one presented above.

NOW IT'S YOUR TURN

(NOTE: Ending a number in 1/2 is equivalent to ending it in .5)

Basic Exercises

1. $28 \times 3.5 =$
2. $18 \times 2.5 =$
3. $6 \times 8.5 =$
4. $9.5 \times 8 =$
5. $1.5 \times 48 =$
6. $7.5 \times 22 =$
7. $6.5 \times 8 =$
8. $18 \times 4\frac{1}{2} =$
9. $106 \times 5\frac{1}{2} =$
10. $14 \times 6\frac{1}{2} =$
11. $6 \times 9\frac{1}{2} =$
12. $24 \times 3\frac{1}{2} =$
13. $32 \times 1\frac{1}{2} =$

Math Masters (Suggestion: Review "Food for Thought" Section.)

14. $800 \times 8.5 =$
15. $90 \times 1.5 =$
16. $6.2 \times 5.5 =$
17. $12 \times 75 =$
18. $1.6 \times 45 =$
19. $250 \times 2.4 =$
20. $0.35 \times 120 =$

(*See answers on page 194*)

DAY 5

TECHNIQUE 18:
Dividing by 1.5, 2.5, 3.5, etc.

BASIC APPLICATION

You've finally decided to build that block wall you've always wanted to have as a backdrop for your garden. The wall will be 18 feet long, and will consist of blocks 1½ (or 1.5) feet long. How many blocks will be needed for each row of your wall?

Calculation: $18 \div 1.5 = ?$

The Old-Fashioned Way

$$
\begin{array}{r}
12 \\
15\overline{)180} \\
\underline{15} \\
30 \\
\underline{30} \\
0
\end{array}
$$

Note: The decimal was removed from the divisor and a zero tacked on to the dividend

The Excell-erated Way

The assumption here is that it is easier to divide by a whole number than by a mixed number (such as 1½ or 1.5). The trick is to double both the dividend and divisor, producing a calculation of only whole numbers. Using the example above, we would double both the 18 and the 1.5, producing the calculation $36 \div 3$. Even though the numbers are now larger, the calculation is easier, as illustrated below.

78

To divide 18 by 1.5, answer the following questions

A. What is 18 × 2?   Answer: 36
B. What is 1.5 × 2?   Answer: 3
C. What is 36 ÷ 3?   Answer: 12 (the solution to 18 ÷ 1.5)

Let's Try One More Basic Calculation:

To divide 27 by 4.5, answer the following questions

A. What is 27 × 2?   Answer: 54
B. What is 4.5 × 2?   Answer: 9
C. What is 54 ÷ 9?   Answer: 6 (the solution to 27 ÷ 4.5)

Math-Master Application

The local weight-loss clinic claims that its patients have lost a combined 440 pounds during the 5½ (or 5.5) months it has been conducting business. How many pounds per month does that average for the patients as a whole?

Calculation: 440 ÷ 5.5 = ?

To divide 440 by 5.5, answer the following questions

A. What is 440 × 2?   Answer: 880
B. What is 5.5 × 2?   Answer: 11
C. What is 880 ÷ 11?   Answer: 80 (or take 88 ÷ 11, and
                             tack on a zero, to obtain the solution to
                             440 ÷ 5.5)

Food for Thought

Even though presentation of this technique has been limited to numbers ending in .5, it works with any number ending in 5. Let's alter the weight-loss example to read, ". . . its 55 patients have lost a combined 440 pounds during the past month. How many pounds does that average per patient?" If you double both the 440 and the 55, you produce a much easier problem: 880 ÷ 110 = 8 pounds per person. Why not try this advanced technique when you become accustomed to the basic applications?

NOW IT'S YOUR TURN

Basic Exercises

1. $22 \div 5.5 =$

2. $340 \div 8.5 =$

3. $390 \div 6.5 =$

4. $375 \div 7.5 =$

5. $140 \div 3.5 =$

6. $10\frac{1}{2} \div 1.5 =$

7. $225 \div 2.5 =$

8. $28\frac{1}{2} \div 9\frac{1}{2} =$

9. $26 \div 6\frac{1}{2} =$

10. $440 \div 5\frac{1}{2} =$

11. $21 \div 1\frac{1}{2} =$

12. $27 \div 4\frac{1}{2} =$

13. $24\frac{1}{2} \div 3\frac{1}{2} =$

Math Masters (Suggestion: Review "Food for Thought" Section.)

14. $180 \div 45 =$

15. $450 \div 75 =$

16. $210 \div 35 =$

17. $195 \div 65 =$

18. $255 \div 15 =$

19. $255 \div 85 =$

20. $38.5 \div 55 =$

(See answers on page 194)

THE HOUSE ALWAYS HAS THE ADVANTAGE

Let's say you're at a carnival, and you spot a dice game. You're asked to bet $1 that, within three rolls of the die, the number you have chosen (1 through 6) will turn up. One would think that with three rolls and six sides to a die, the odds would be 50/50. Not so! The odds are $1 - (\%)^3$, which equals about 42 chances in 100.

DAY 5

TECHNIQUE 19:
Squaring Any Number Ending in 1

BASIC APPLICATION

You're at the doctor's office, lying on your back in the examination room. While waiting for the doctor, you start counting the little holes in the ceiling tile. You notice that each square tile contains 31 holes to a side. How many holes are there per tile?

Calculation: $31 \times 31 = ?$

The Old-Fashioned Way

$$\begin{array}{r} 31 \\ \times\, 31 \\ \hline 31 \\ 93 \\ \hline 961 \end{array}$$

The Excell-erated Way

The theory behind this technique is that it's easier to square a number ending in 0 than it is to square a number ending in 1. That is, we begin by squaring the number, less 1, and go from there. Using the example above, $31^2 = 30^2 + 30 + 31$. If, instead, we were squaring 81, we would add $80^2 + 80 + 81$. Let's see this nifty little trick in action.

To square the number 31, answer the following questions

A. What is the whole number before 31?   Answer: 30

B. What is 30^2 (or 30×30)   Answer: 900

C. What is $900 + 30 + 31$?   Answer: 961 (the solution to 31^2).

(NOTE: To square 30, simply square 3, and tack on *two* zeroes)

Let's Try One More Basic Calculation: 61×61

To square the number 61, answer the following questions

A. What is the whole number before 61?   Answer: 60

B. What is 60^2?   Answer: 3,600

C. What is $3,600 + 60 + 61$?   Answer: 3,721 (the solution to 61^2)

Math-Master Application

The hotel you manage has a vacancy rate of 21 percent. If there are 2,100 rooms in all, how many rooms are vacant?

Calculation: $2,100 \times 21\%$

              $(2,100 \times 0.21)$

To multiply 2,100 by 21%, answer the following questions

A. What does the calculation look like, omitting the zeroes and percentage symbol?   Answer: 21×21

B. What is the whole number before 21?   Answer: 20

C. What is 20^2?   Answer: 400

D. What is $400 + 20 + 21$?   Answer: 441 (tentative solution)

Now let's test the tentative solution, 441, to see if it appears to be the solution to $2,100 \times 21\%$. We know that 10% (or $\frac{1}{10}$) of 2,100 is 210, so 21% of 2,100 should be a bit more than 420 (twice 210). Therefore, the tentative solution is also the final solution.

Food for Thought

This technique will actually work with any square. For example, you could square 26 by also using Technique #14 and taking $25^2 + 25 + 26$, which equals 676. Applying Technique #19 in reverse, you could square 19 by taking $20^2 - 20 - 19$, which equals 361. These variations, however, take a significant amount of concentration to apply. For that reason, you may be better off using this technique only to square numbers ending in 1.

NOW IT'S YOUR TURN

Basic Exercises

1. $61 \times 61 =$
2. $21 \times 21 =$
3. $51 \times 51 =$
4. $11^2 =$
5. $91^2 =$
6. $41^2 =$
7. $71 \times 71 =$
8. $31 \times 31 =$
9. $81 \times 81 =$
10. $21^2 =$
11. $61^2 =$
12. $51^2 =$
13. $41 \times 41 =$

Math Masters

14. $9.1 \times 9.1 =$
15. $3.1 \times 310 =$

16. $0.81 \times 81 =$

17. $710^2 =$

18. $210 \times 2.1 =$

19. $41 \times 4.1 =$

20. $5.1 \times 5.1 =$

(See answers on page 194)

DAY 5

TECHNIQUE 20:
Multiplying Two Numbers Whose Difference Is 2

BASIC APPLICATION

You are about to mail a package that weighs exactly 14 pounds. Given that there are 16 ounces to a pound, how many ounces are there in 14 pounds?

Calculation: $14 \times 16 = ?$

The Old-Fashioned Way   The Excell-erated Way

$$\begin{array}{r} \overset{2}{14} \\ \times\ 16 \\ \hline 84 \\ 14 \\ \hline 224 \end{array}$$

This technique applies to calculations such as 9×11, 17×15, and 94×96. All you have to do is square the number in the middle, and subtract 1. Using the example above, $14 \times 16 = 15^2 - 1$. Of course, application of this technique depends upon your ability to square numbers. Numbers ending in zero, such as 20 or 70, are easy to square, as are numbers ending in 5 (use Technique #14) and two-digit numbers ending in 1 (use Technique #19). You might want to review the squares presented on page xiv. Let's see how this interesting trick works.

To multiply 14 by 16, answer the following questions

A. What is the number in the middle?   Answer: 15

B. What is 15^2?   Answer: 225

C. What is $225 - 1$?   Answer: 224 (the solution to 14×16)

Let's Try One More Basic Calculation: 31×29

To multiply 31 by 29, answer the following questions

A. What is the number in the middle?   Answer: 30

B. What is 30^2?   Answer: 900

C. What is $900 - 1$?   Answer: 899 (the solution to 31×29)

Math-Master Application

Let's assume the price of gasoline is currently $1.30 per gallon. You expect that within five years, gasoline prices will jump by 50 percent. If your prediction comes true, what will the price of gasoline be five years from now?

Calculation: $1.30 $\times$ 1.5 = ?
              ($1.30 $\times$ 150%)

(You'll note that 1.3 and 1.5 have a difference of 0.2, but this technique will still work because 13 and 15 have a difference of 2.)

To multiply $1.30 by 1.5, answer the following questions

A. What does the calculation look like, omitting the decimal points, zero, and dollar sign?   Answer: 13×15

B. What is the number in the   Answer: 14
    middle?
C. What is 14^2?   Answer: 196
D. What is $196 - 1$?   Answer: 195, or $195 (tentative
                                        solution)

Now let's test the tentative solution, $195, to see if it appears to be the solution to $1.30 × 1.5. By inspection, we can see that $195 is much too large, $19.50 is still too large, but $1.95 fits the bill perfectly. In each case, we moved the decimal point one place to the left until we found the solution.

Food for Thought

There are an infinite number of additional techniques that work just like this one. For example, to multiply two numbers with a difference of 4, square the number in the middle and subtract 4 (e.g., $8 × 12 = 10^2 - 4 = 96$. Can you figure out how to apply this technique to two numbers with a difference of 10, such as 15 × 25? That's right, simply square the number in the middle (20, in this case) and subtract 25, producing (in this case) the answer 375. Experiment, and use the ones that most tickle your fancy!

NOW IT'S YOUR TURN

Basic Exercises

1. $19 × 21 =$

2. $11 × 13 =$

3. $46 × 44 =$

4. $91 × 89 =$

5. $14 × 12 =$

6. $74 × 76 =$

7. $49 × 51 =$

8. $13 × 15 =$

9. $26 \times 24 =$

10. $101 \times 99 =$

11. $17 \times 15 =$

12. $64 \times 66 =$

13. $79 \times 81 =$

Math Masters

14. $3.6 \times 34 =$

15. $14 \times 1.6 =$

16. $7.1 \times 6.9 =$

17. $96 \times 0.94 =$

18. $120 \times 1.4 =$

19. $39 \times 4.1 =$

20. $5.4 \times 56 =$

(See answers on page 194)

NUMBER CURIOSITY

$$1 = \quad 1 = 1^3$$
$$3 + 5 = \quad 8 = 2^3$$
$$7 + 9 + 11 = \quad 27 = 3^3$$
$$13 + 15 + 17 + 19 = \quad 64 = 4^3$$
$$21 + 23 + 25 + 27 + 29 = 125 = 5^3$$
$$\text{etc.}$$

DAY 5 REVIEW QUIZ

DIRECTIONS: Work each exercise as quickly as possible, using the techniques just learned.

Basic Exercises

1. $3\frac{1}{2} \times 24 =$ 5. $24 \times 25 =$

2. $27 \div 4.5 =$ 6. $81^2 =$

3. $31 \times 31 =$ 7. $39 \div 6\frac{1}{2} =$

4. $51 \times 49 =$ 8. $4.5 \times 18 =$

Math Masters

9. $45 \times 1.6 =$ 11. $210 \div 35 =$

10. $6.1 \times 6.1 =$ 12. $21 \times 1.9 =$

(See answers on page 195)

QUICK CHALLENGE

If a piece of land is 40 yards square, how many yards square is ¼ of the piece of land?

ANSWER: 20 yards square (not 10 yards square!)

MIDTERM EXAM

DIRECTIONS: Work each exercise as quickly as possible, using the techniques learned.

Basic Exercises

1. $5 + 8 + 2 + 1 + 4 + 9 + 3 =$
2. $32 + 19 + 8 + 40 + 15 + 24 =$
3. $14 + 3 + 25 + 33 + 42 + 7 =$
4. $39 + 75 =$
5. $41 - 17 =$
6. $121 - 84 =$
7. $67 - 39 =$
8. $140 - 83 =$
9. $52 \times 4 =$
10. $220 \div 4 =$
11. $5 \times 72 =$
12. $22 \div 5 =$
13. $62 \times 11 =$
14. $35^2 =$
15. $68 \times 25 =$
16. $320 \div 25 =$
17. $5.5 \times 26 =$
18. $21 \div 1.5 =$
19. $41^2 =$
20. $31 \times 29 =$

Math Masters

21. $67 + 24 + 32 + 88 + 12 =$

22. $19 + 30 + 47 + 88 + 6 =$

23. $59 + 115 =$

24. $150 - 73 =$

25. $606 - 558 =$

26. $400 \times 0.35 =$

27. $380 \div 50 =$

28. $1.1 \times 220 =$

29. $115 \times 115 =$

30. $1,700 \div 250 =$

31. $180 \div 45 =$

32. $1.6 \times 140 =$

(*See answers on page 195*)

DAY 6:

MULTIPLICATION AND DIVISION IV

DAY 6

TECHNIQUE 21:
Multiplying by 15

BASIC APPLICATION

Your car gets exactly 15 miles per gallon and has an 18-gallon gas tank. How many miles would you expect to drive on a full tank before running out of gas?

Calculation: $18 \times 15 = ?$

The Old-Fashioned Way

$$
\begin{array}{r}
\overset{4}{18} \\
\times\ 15 \\
\hline
90 \\
18 \\
\hline
270
\end{array}
$$

The Excell-erated Way

Multiplying a number by 15 is really the same as multiplying the number by 10 (just tack on a zero), then multiplying the number by 5 (take half the product just obtained), and adding the two products. Using the example above, $18 \times 15 = (18 \times 10) + \dfrac{(18 \times 10)}{2}$

Read on for a step-by-step explanation.

To multiply 18 by 15, answer the following questions

A. What is 18 × 10?   Answer: 180
B. What is half of 180?   Answer: 90
C. What is 180 + 90?   Answer: 270 (the solution to 18 × 15)

Let's Try One More Basic Calculation: 15 × 24

To multiply 15 by 24, answer the following questions

A. What is 24 × 10?   Answer: 240
B. What is half of 240?   Answer: 120
C. What is 240 + 120?   Answer: 360 (the solution to 15 × 24)

Math-Master Application

This technique will also work when multiplying a number by 0.15, 1.5, 150, 1,500, and so on. Let's say you're in a restaurant with friends, and you've volunteered to leave the tip. The check comes to $34.00, and you'd like to leave 15% for the waiter. How much of a tip should you leave?

Calculation: $34.00 × 15% = ?

(NOTE: $34.00 × 15% is the same as $34.00 × 0.15)

To multiply $34.00 by 15%, answer the following questions

A. What does the calculation look like,   Answer: 34 × 15
    omitting the dollar sign, zeroes, decimal
    point, and percentage symbol?
B. What is half of 34?   Answer: 17
C. What is 34 + 17?   Answer: 51
D. Where should the decimal point now be placed?

    Answer: Well, $51.00 would be more than the check itself, and
    $.51 (51 cents) appears ridiculously low. But $5.10 sounds just
    right, and in fact is the answer!

Food for Thought

A variation on the above technique to multiply a number by 15 is to *first* add half of the number and *then* multiply by 10. For example, $18 \times 15 = (18 + 9) \times 10 = 270$. This variation will produce a fraction, however, and may be awkward to apply when multiplying an odd number by 15. For example, $17 \times 15 = (17 + 8\frac{1}{2}) \times 10 = 255$. Use whichever variation you find easier to apply.

NOW IT'S YOUR TURN

Basic Exercises

1. $32 \times 15 =$
2. $14 \times 15 =$
3. $22 \times 15 =$
4. $15 \times 36 =$
5. $15 \times 16 =$
6. $15 \times 28 =$
7. $44 \times 15 =$
8. $26 \times 15 =$
9. $38 \times 15 =$
10. $15 \times 64 =$
11. $15 \times 52 =$
12. $15 \times 34 =$
13. $66 \times 15 =$

Math Masters

14. $4.6 \times 150 =$
15. $62 \times 1.5 =$

16. $15 \times 72 =$

17. $1.5 \times 180 =$

18. $150 \times 80 =$

19. $3.2 \times 150 =$

20. $84 \times 1.5 =$

(See answers on page 195)

DAY 6

TECHNIQUE 22:
Multiplying Two Numbers Just Over 100

BASIC APPLICATION

You've just purchased a piece of land that measures 105 feet by 102 feet. How many square feet have you bought?

Calculation: $105 \times 102 = ?$

The Old-Fashioned Way

```
    105
 ×  102
 ------
    210
  10 50
 ------
 10,710
```

The Excell-erated Way

This technique will work for calculations such as 107×103, 102×126, and 109×111 (see the "Food for Thought" section for the upper limit). The solution will always be a five-digit number beginning in 1. Using the example above, the next two digits will be the *sum* of the 5 and the 2, and the last two digits will be the *product* of the 5 and the 2. If either the sum or the product consists of one digit (as 7), a zero is placed in front (producing 07 in this case). Let's examine this fascinating trick in a step-by-step fashion.

To multiply 105 by 102, answer the following questions

A. How will the answer   Answer: 1
    automatically begin?
B. What is 5 + 2?   Answer: 07 (note attached zero)
C. What is 5 × 2?   Answer: 10
D. What do the combined   Answer: 10,710 (solution)
    answers produce?

Let's Try One More Basic Calculation: 104 × 107

To multiply 104 by 107, answer the following questions

A. How will the answer   Answer: 1
    automatically begin?
B. What is 4 + 7?   Answer: 11
C. What is 4 × 7?   Answer: 28
D. What do the combined   Answer: 11,128 (solution)
    answers produce?

Math-Master Application

You're a college professor, and on the first day of class you always give your students a set of handouts to cover the entire course. You have 103 students and each set of handouts consists of 122 pages. How many pages will you have to photocopy in all?

Calculation: 103 × 122 = ?

To multiply 103 by 122, answer the following questions

A. How will the answer   Answer: 1
    automatically begin?
B. What is 3 + 22?   Answer: 25
C. What is 3 × 22?   Answer: 66
D. What do the combined   Answer: 12,566 (solution)
    answers produce?

Food for Thought

This technique will also work with decimals, such as $1.05 × 1,070. However, it will not work if the multiplication step produces a product greater than 99. For example, 104 × 132 will not work (at least not without carrying) because 4 × 32 equals more than 99. On the other hand, 109 × 111 will work because 9 × 11 does not exceed 99. You can practice this technique when you notice posted gasoline prices, which in recent years have slightly exceeded $1.00 per gallon.

NOW IT'S YOUR TURN

Basic Exercises

1. 102 × 107 =
2. 109 × 109 =
3. 107 × 101 =
4. 104 × 104 =
5. 108 × 107 =
6. 101 × 109 =
7. 106 × 104 =
8. 102 × 101 =
9. 103 × 105 =
10. 105 × 108 =
11. 107 × 109 =
12. 101 × 101 =
13. 103 × 102 =

Math Masters

14. 105 × 115 =
15. 121 × 104 =

16. $111 \times 109 =$

17. $103 \times 132 =$

18. $106 \times 112 =$

19. $145 \times 102 =$

20. $101 \times 198 =$

(*See answers on page 195*)

FOR YOUR INFORMATION

What is a prime number? It's a number, such as 3, 11, or 29, that cannot be divided evenly by any whole number other than itself and 1. What's the highest known prime number? It's a number so large it contains over 65,000 digits!

DAY 6

TECHNIQUE 23:
Multiplying by 75

BASIC APPLICATION

Your dining room window measures 52 inches by 75 inches. How many square inches of wall space does the window cover?

Calculation:

$$52 \times 75 = ?$$

The Old-Fashioned Way

$$
\begin{array}{r}
{\scriptstyle 1\ 1} \\
52 \\
\times\ 75 \\
\hline
260 \\
3\ 64 \\
\hline
3{,}900
\end{array}
$$

The Excell-erated Way

Even though this is a calculation of whole numbers, you can solve it by thinking "¾" whenever you see the number 75. That is, to multiply a number by 75, take ¾ of the number, and then multiply by 100 to obtain the answer. The best way to multiply a number by ¾ is to take half of the number, and then add half again. For example, ¾ of 16 equals half of 16 (or 8) plus half of 8 (or 4), which equals 12. Let's try this technique with our window example.

To multiply 52 by 75, answer the following questions

A.   What is ¾ of 52?   Answer: 39 (that is, 26 plus 13)
B.   What is 39 × 100?   Answer: 3,900 (the solution to
                                          52 × 75)

Let's Try One More Basic Calculation: 32 × 75

To multiply 32 by 75, answer the following questions

A.   What is ¾ of 32?   Answer: 24 (that is, 16 plus 8)
B.   What is 24 × 100?   Answer: 2,400 (the solution to
                                          32 × 75)

Math-Master Application

You've had your eye on a certain fax machine to purchase, but are waiting for a sale. Finally, you see the machine advertised at 25 percent off the suggested retail price of $480. What is the sale price of the fax machine? (Hint: Most people will figure 25% of $480, and subtract that amount from $480. However, you can obtain the answer in one step by simply multiplying $480 by 75%. Why will this work? Because subtracting 25% is identical with multiplying by 75%).

Calculation: $480 × 75% = ?

To multiply $480 by 75%, answer the following questions

A.   What does the calculation   Answer: 48 × 75
     look like, omitting the dollar
     sign, zero, and percentage
     symbol?
B.   What is 48 × ¾?   Answer: $36 (that is, 24 plus 12)
                                             (this is the tentative solution)

Because we've altered the original calculation, there is no need at this point to multiply the tentative solution, $36, by 100. Instead, we test it to see if it appears to be the solution to $480 × 75%. It should be obvious that $36 is too small, and $3,600 is impossibly large, to be the solution. However, $360 is just right!

Food for Thought

Perhaps you've already figured it out, but to multiply a number by 75, you could first multiply the number by 100, and *then* take ¾. Taking the original calculation above, 52×75, you could first multiply 52 by 100, obtaining 5,200. Then you would take ¾, which equals 3,900. Normally, however, it's easier to first take ¾, and *then* multiply by 100 (calculating ¾ of a smaller number is easier even if the larger number has trailing zeroes). In fact, if you want, you can forget about performing the multiplication by 100 altogether. Simply take ¾, and test the tentative solution for zeroes to tack on.

NOW IT'S YOUR TURN

Basic Exercises

1. $24 \times 75 =$
2. $12 \times 75 =$
3. $36 \times 75 =$
4. $75 \times 44 =$
5. $75 \times 8 =$
6. $75 \times 28 =$
7. $48 \times 75 =$
8. $56 \times 75 =$
9. $72 \times 75 =$
10. $75 \times 16 =$
11. $75 \times 88 =$
12. $75 \times 68 =$
13. $64 \times 75 =$

Math Masters

14. $1.2 \times 750 =$

15. $36 \times 7.5 =$

16. $0.75 \times 440 =$

17. $18 \times 75 =$

18. $2.4 \times 7.5 =$

19. $26 \times 75 =$

20. $7.5 \times 140 =$

(See answers on page 195)

DAY 6

TECHNIQUE 24:
Dividing by 75

BASIC APPLICATION

You've purchased a book that will help you prepare for the professional examination you're scheduled to take next month. After some practice, you find you can answer 75 questions in 180 minutes. What is your average time to complete one item?

Calculation: $180 \div 75 = ?$

The Old-Fashioned Way

```
      2.4
75)180.0
   150
    30 0
    30 0
       0
```

The Excell-erated Way

This is another technique that makes use of fractions. To divide a number by 75, multiply the number by $1\frac{1}{3}$, and then divide by 100. The best way to multiply by $1\frac{1}{3}$ is to take $\frac{1}{3}$ of the number and add it to the number. For example, $1\frac{1}{3}$ of 12 equals $\frac{1}{3}$ of 12, which is 4, plus 12, equaling 16. Let's see how this technique will solve the problem stated above.

To divide 180 by 75, answer the following questions

A.   What is 1⅓ of 180?   Answer: 240 (that is, 60 plus 180)
B.   What is 240 ÷ 100?   Answer: 2.4 (the solution to 180 ÷ 75)

(NOTE: In "A" above, you could have taken 1⅓ of 18 [or 24], and then tacked on a zero, producing the same 240.)

Let's Try One More Basic Calculation:

To divide 900 by 75, answer the following questions

A.   What is 1⅓ of 900?   Answer: 1,200 (that is, 300 plus 900)
B.   What is 1,200 ÷ 100?   Answer: 12 (the solution to 900 ÷ 75)

(NOTE: Here, again, you could have ignored the zeroes, and simply taken 1⅓ of 9 in "A" above, to produce the answer 12.)

Math-Master Application

During his professional career, Hank Aaron hit approximately 750 home runs. In addition, he came to bat approximately 12,000 times. He, therefore, hit a round-tripper about once in every how many at-bats?

Calculation: 12,000 ÷ 750 = ?

To divide 12,000 by 750, answer the following questions

A.   What does the calculation   Answer: 12 ÷ 75
     look like, omitting all the
     zeroes?
B.   What is 12 × 1⅓?   Answer: 16 (tentative solution)

Let's test the tentative solution to see if it appears to be the solution to 12,000 ÷ 750. Eliminating one zero each from the dividend and divisor, we get 1,200 ÷ 75. Then, working backwards, we ask ourselves if 75 × 16 = 1,200. We can readily see that 75 × 1.6 wouldn't nearly equal 1,200, and 75 × 160 equals far more than 1,200. Therefore, 75 × 16 seems to fit the bill, with 16 being the correct answer.

Food for Thought

Here is why this trick works. Remember from Technique #23 that you should think "¾" when multiplying by 75. In this case, however, we are dividing by 75. Dividing by ¾ is the same as multiplying by the inverse, which is ⁴⁄₃, or 1⅓. Now you may be thinking that learning to multiply by 1⅓ will take some practice. However, hopefully you'll agree that this technique is still faster and easier than the conventional way of dividing by 75.

NOW IT'S YOUR TURN

Basic Exercises

1. $120 \div 75 =$

2. $210 \div 75 =$

3. $3,600 \div 75 =$

4. $270 \div 75 =$

5. $450 \div 75 =$

6. $390 \div 75 =$

7. $6,000 \div 75 =$

8. $3,300 \div 75 =$

9. $510 \div 75 =$

10. $990 \div 75 =$

11. $180 \div 75 =$

12. $4,200 \div 75 =$

13. $660 \div 75 =$

Math Masters

14. $7,200 \div 75 =$

15. $930 \div 75 =$

16. $5,400 \div 750 =$

17. $8.1 \div 7.5 =$

18. $8,700 \div 75 =$

19. $630 \div 7.5 =$

20. $12,000 \div 750 =$

(See answers on page 195)

NUMBER CURIOSITY

$$65,359,477,124,183 \times 17 = 1,111,111,111,111,111$$
$$65,359,477,124,183 \times 34 = 2,222,222,222,222,222$$
$$65,359,477,124,183 \times 51 = 3,333,333,333,333,333$$
$$65,359,477,124,183 \times 68 = 4,444,444,444,444,444$$
$$65,359,477,124,183 \times 85 = 5,555,555,555,555,555$$

etc.

DAY 6 REVIEW QUIZ

DIRECTIONS: Work each exercise as quickly as possible, using the techniques just learned.

Basic Exercises

1. $52 \times 15 =$

2. $108 \times 103 =$

3. $16 \times 75 =$

4. $450 \div 75 =$

5. $1,200 \div 75 =$

6. $75 \times 56 =$

7. $105 \times 107 =$

8. $15 \times 44 =$

Math Masters

9. $1.5 \times 320 =$

11. $7.5 \times 24 =$

10. $133 \times 102 =$

12. $6,300 \div 750 =$

(See answers on page 196)

QUICK TEASER

Can you figure out a technique to quickly add the whole numbers from 1 to n? For example, how would you quickly calculate $1 + 2 + 3 + 4 + 5$?

ANSWER: Take the top number (5, in this case), multiply by the next whole number (6, in this case), and divide the product (30, in this case) by 2 (producing the answer 15, in this case). The formula, therefore, is $\dfrac{n \times (n + 1)}{2}$, where "n" is the highest number in the series.

DAY 7:

MULTIPLICATION AND DIVISION V

DAY 7

TECHNIQUE 25:
Multiplying by 9

BASIC APPLICATION

You've just purchased 18 square yards of material to make some curtains. You'd rather deal in square feet, since you've measured your windows in feet. There are 9 square feet to a square yard, so how many square feet are there in 18 square yards of material?

Calculation: $18 \times 9 = ?$

The Old-Fashioned Way

```
  7
 18
× 9
───
162
```

The Excell-erated Way

The theory behind this technique is that it's easier to multiply by 10 than it is to multiply by 9. Specifically, multiplying a number by 9 is the same as multiplying by 10 and then subtracting the number. Using the example above, $18 \times 9 = (18 \times 10) - 18$. Let's see how this commonly used technique works.

To multiply 18 by 9, answer the following questions

A. What is 18×10?   Answer: 180
B. What is $180 - 18$?   Answer: 162 (the solution to 18×9)

115

Let's Try One More Basic Calculation: 33 × 9

To multiply 33 by 9, answer the following questions

A. What is 33 × 10?   Answer: 330
B. What is 330 − 33?   Answer: 297 (the solution to 33 × 9)

Math-Master Application

If your department were to work at 100% efficiency, it could produce 2,400 widgets per day. Yesterday, operations proceeded at 90% efficiency. How many widgets, therefore, were produced?

Calculation: 2,400 × 90%
              (or 2,400 × 0.9)

To multiply 2,400 by 0.9, answer the following questions

A. What does the calculation   Answer: 24 × 9
    look like, omitting the zeroes
    and decimal point?
B. What is 24 × 10?   Answer: 240
C. What is 240 − 24?   Answer: 216 (tentative solution)

Let's test the tentative solution by reexamining the original calculation, 2,400 × 0.9. Since 0.9 is slightly less than 1, the solution must be a bit less than 2,400. Well, the tentative solution, 216, is more than a bit less than 2,400. Tacking on a zero, however, will produce the number we're looking for—2,160.

Food for Thought

The good news about this technique is that we begin with a simple calculation—multiplying by 10. The bad news, however, is that we must then perform the dreaded subtraction. You may wish to brush up on the subtraction techniques presented during Day 2 of this program. Going back to the "Math Master" application above, one way to subtract 24 from 240 is to first subtract 20 from 240, producing 220, and then subtract 4, producing 216.

NOW IT'S YOUR TURN

Basic Exercises

1. $12 \times 9 =$
2. $19 \times 9 =$
3. $24 \times 9 =$
4. $9 \times 14 =$
5. $9 \times 26 =$
6. $9 \times 16 =$
7. $45 \times 9 =$
8. $17 \times 9 =$
9. $15 \times 9 =$
10. $9 \times 28 =$
11. $9 \times 35 =$
12. $9 \times 23 =$
13. $25 \times 9 =$

Math Masters

14. $130 \times 9 =$
15. $4.7 \times 90 =$
16. $0.9 \times 38 =$
17. $90 \times 34 =$
18. $9 \times 5.8 =$
19. $0.89 \times 900 =$
20. $6.7 \times 90 =$

(See answers on page 196)

DAY 7

TECHNIQUE 26:
Multiplying by 12

BASIC APPLICATION

The doughnut shop you manage has sold 31 dozen doughnuts during its first hour of operations today. How many doughnuts have you sold?

Calculation: $31 \times 12 = ?$

The Old-Fashioned Way

$$\begin{array}{r} 31 \\ \times\ 12 \\ \hline 62 \\ 31 \\ \hline 372 \end{array}$$

The Excell-erated Way

Multiplying by 12 is similar to multiplying by 15 (Technique #21) in that you divide the calculation into two parts. Specifically, you first multiply the number by 10, then multiply it by 2 (i.e., double it), and then add the two products. Using the example above, $31 \times 12 = (31 \times 10) + (31 \times 2)$. Let's see this calculation in step-by-step form.

To multiply 31 by 12, answer the following questions

A. What is 31 × 10? Answer: 310
B. What is 31 × 2? Answer: 62
C. What is 310 + 62? Answer: 372 (the solution to 31 × 12)

Let's Try One More Basic Calculation: 24 × 12

To multiply 24 by 12, answer the following questions

A. What is 24 × 10? Answer: 240
B. What is 24 × 2? Answer: 48
C. What is 240 + 48? Answer: 288 (the solution to 24 × 12)

Math-Master Application

You've just opened a review course to help people prepare for the CPA exam. As a promotional gimmick, you've purchased 750 pencils, printed with your course's name and phone number, to distribute to your students. If each pencil cost you 12 cents, how much did you pay for the 750?

Calculation: 750 × 12¢ = ?
(750 × $.12)

To multiply 750 by $.12, answer the following questions

A. What does the calculation look like, omitting the zero, dollar sign, and decimal point? Answer: 75 × 12
B. What is 75 × 10? Answer: 750
C. What is 75 × 2? Answer: 150
D. What is 750 + 150? Answer: 900, or $900 (tentative solution)

Let's test the tentative solution. We know that 12 cents is slightly more than 10% (or 1/10) of a dollar. Therefore, the solution to 750 × $.12 should be a bit more than 1/10 of 750, or $75. The tentative solution, $900, is impossibly high, whereas $90 fills the bill as being slightly more than $75.

Food for Thought

As was stated in an earlier technique, there is often more than one way to test a tentative solution to see if it is the correct answer. Similarly, there is often more than one way to approach a calculation. Using the pencil example above, can you think of another way to perform the computation? That's right, you can use Technique #23 (Multiplying by 75). All you have to do is take ¾ of 12 (or 9), attach a zero and dollar sign, and you've got the answer, $90!

NOW IT'S YOUR TURN

Basic Exercises

1. $14 \times 12 =$

2. $33 \times 12 =$

3. $15 \times 12 =$

4. $12 \times 35 =$

5. $12 \times 18 =$

6. $12 \times 22 =$

7. $55 \times 12 =$

8. $17 \times 12 =$

9. $75 \times 12 =$

10. $12 \times 16 =$

11. $12 \times 23 =$

12. $12 \times 25 =$

13. $32 \times 12 =$

Math Masters

14. $140 \times 1.2 =$

15. $19 \times 120 =$

16. $12 \times 6.5 =$

17. $120 \times 0.85 =$

18. $1.2 \times 210 =$

19. $4.5 \times 120 =$

20. $3.4 \times 1.2 =$

(See answers on page 196)

NUMBER CURIOSITY

$$15{,}873 \times 7 = 111{,}111$$
$$15{,}873 \times 14 = 222{,}222$$
$$15{,}873 \times 21 = 333{,}333$$
$$15{,}873 \times 28 = 444{,}444$$
$$15{,}873 \times 35 = 555{,}555$$
etc.

DAY 7

TECHNIQUE 27:
Multiplying by 125

BASIC APPLICATION

You're a college professor, and you've just given your 125 students a 24-question exam. How many questions, in all, will you (or your teaching assistant) have to grade?

Calculation: $24 \times 125 = ?$

The Old-Fashioned Way

$$
\begin{array}{r}
1 \\
12 \\
125 \\
\times\ 24 \\
\hline
500 \\
2\ 50 \\
\hline
3,000
\end{array}
$$

The Excell-erated Way

Even though the number 125 is relatively large, it is commonly involved in calculations because it is a multiple of 25. Multiplying a number by 125 is identical with dividing the number by 8, and then multiplying by 1,000 (i.e., tacking on three zeroes). In the example above, $24 \times 125 = (24 \div 8) \times 1,000$. As you'll see, this technique is very similar to Techniques #11 and #15.

To multiply 24 by 125, answer the following questions

A. What is 24 ÷ 8?   Answer: 3
B. What is 3 × 1,000?   Answer: 3,000 (the solution to
                                24 × 125)

Let's Try One More Basic Calculation: 56 × 125

To multiply 56 by 125, answer the following questions

A. What is 56 ÷ 8?   Answer: 7
B. What is 7 × 1,000?   Answer: 7,000 (the solution to
                                56 × 125)

Math-Master Application

At the beginning of this month, the balance on your personal line of credit was $3,200. The annual interest rate on your account is 15%, which translates to 1¼% (or 1.25%) per month. At the end of this month, how much interest will you have incurred for the month?

Calculation: $3,200 × 1.25%

(NOTE: 1.25% may be expressed as 0.0125)

To multiply $3,200 by 1.25%, answer the following questions

A. What does the calculation   Answer: 32 × 125
    look like, ignoring the dollar
    sign, zeroes, decimal point,
    and percent symbol?
B. What is 32 ÷ 8?   Answer: 4 (tentative solution)

Let's test the tentative solution by reexamining the original calculation. Since 1.25% is slightly more than 1%, then the correct answer must be a bit greater than 1% × $3,200, or $32. Tacking a zero onto the tentative solution will produce the correct answer, $40.

Food for Thought

If the numbers comprising a calculation don't quite fit the description for a trick, but all you need is an estimate, you can still use the trick. Using Technique #27 as an example, if the calculation at hand is 24 × 124 or

24×126, you can pretend that it is 24×125 and you'll be within 1 percent of the correct answer. If you don't mind being about 1.6 percent off, you can use this estimation technique for multiplication by 123 and 127. Be careful not to stray too far, however, from the technique's specifications if you want a decent estimate.

NOW IT'S YOUR TURN

Basic Exercises

1. $32 \times 125 =$
2. $64 \times 125 =$
3. $48 \times 125 =$
4. $125 \times 72 =$
5. $125 \times 16 =$
6. $125 \times 88 =$
7. $96 \times 125 =$
8. $80 \times 125 =$
9. $560 \times 125 =$
10. $125 \times 24 =$
11. $125 \times 40 =$
12. $125 \times 8 =$
13. $160 \times 125 =$

Math Masters

14. $12 \times 125 =$
15. $36 \times 125 =$
16. $125 \times 60 =$
17. $1.25 \times 240 =$
18. $12.5 \times 6.4 =$

19. $320 \times 1.25 =$

20. $0.88 \times 1,250 =$

(See answers on page 196)

DAY 7

TECHNIQUE 28:
Dividing by 125

BASIC APPLICATION

You've instructed your stockbroker to purchase $8,000 worth of stock for you that she feels has good growth potential. You later learn that she has done so, for $125 per share. How many shares have you, therefore, purchased?

Calculation: $8,000 ÷ $125 = ?

The Old-Fashioned Way

$$\begin{array}{r} 64 \\ 125\overline{)8000} \\ 750 \\ \hline 500 \\ 500 \\ \hline 0 \end{array}$$

The Excell-erated Way

As you may have already guessed, dividing a number by 125 is the same as multiplying the number by 8, and then dividing by 1,000. In most cases it's probably easier to first divide by 1,000, and then multiply by 8. Using the example above, $8,000 ÷ $125 = ($8,000 ÷ 1,000) × 8. The theory here is that it is much easier to multiply by 8 than it is to divide by 125. Let's now practice a few.

To divide $8,000 by $125, answer the following questions

A. What is $8,000 ÷ 1,000?   Answer: $8
B. What is $8 × 8?   Answer: $64 (solution to
                             $8,000 ÷ $125)

(NOTE: Ignoring the three zeroes of $8,000 and skipping right to step "B" is identical with dividing by 1,000.)

Let's Try One More Basic Calculation: 700 ÷ 125

To divide 700 by 125, answer the following questions

A. What is 700 ÷ 1,000?   Answer: 0.7
B. What is 0.7 × 8?   Answer: 5.6 (solution to 700 ÷ 125)

(SUGGESTION: Ignore the zeroes and just multiply 7 by 8, followed by a moment's thought about where to place the decimal point.)

Math-Master Application

Your department has a budget of $450 to spend on computer disks. If each double-box of disks costs $12.50, how many boxes can you purchase with the $450 allotted?

Calculation: $450 ÷ $12.50 = ?

To divide $450 by $12.50, answer the following questions

A. What does the calculation   Answer: 45 ÷ 125
    look like, ignoring the dollar
    signs, zeroes, and decimal
    point?
B. What is 45 × 8?   Answer: 360 (tentative solution)

Let's test the tentative solution. One way is to pretend the original calculation is $450 ÷ $10, because $10 is close to $12.50. The solution to this pretended calculation is 45. However, because $12.50 (the denominator) is greater than $10, the correct answer must be a bit smaller than 45. Lopping off the zero from the tentative solution, we obtain the correct answer, 36.

Food for Thought

You'll note that in the Math-Master calculation above, we simply ignored the division by 1,000, but carefully tested the tentative solution. This is actually the best way to attack all division-by-125 problems. Also, what's the easiest way to multiply a two-digit number (as 45 above) by 8? If you take Technique #9 one step further, double the number, double once again, and double one final time (so think, "45, 90, 180, 360").

NOW IT'S YOUR TURN

Basic Exercises

1. $1,100 \div 125 =$
2. $3,000 \div 125 =$
3. $12,000 \div 125 =$
4. $2,000 \div 125 =$
5. $11,000 \div 125 =$
6. $6,000 \div 125 =$
7. $900 \div 125 =$
8. $1,500 \div 125 =$
9. $4,000 \div 125 =$
10. $2,200 \div 125 =$
11. $500 \div 125 =$
12. $800 \div 125 =$
13. $7,000 \div 125 =$

Math Masters

14. $60 \div 12.5 =$

15. $4 \div 1.25 =$

16. $350 \div 125 =$

17. $11,100 \div 125 =$

18. $70 \div 12.5 =$

19. $2 \div 0.125 =$

20. $50 \div 1.25 =$

(See answers on page 196)

JUST FOR FUN

Let's say you've positioned nine people in front of a piano (enough to cover all 88 keys), and you've asked them to attempt to play all possible sounds from the piano's 88 keys. They can collectively play any number of notes, from only 1 to all 88. Assuming they can (remarkably) play one different sound per second, how many seconds will it take them to accomplish all possible combinations?

ANSWER: $2^{88} - 1$ seconds, or approximately 9.8 quintillion years. (That's about 700 million times the estimated age of the universe!)

DAY 7 REVIEW QUIZ

DIRECTIONS: Work each exercise as quickly as possible, using the techniques just learned.

Basic Exercises

1. $35 \times 9 =$

2. $25 \times 12 =$

3. $24 \times 125 =$

4. $700 \div 125 =$

5. $9,000 \div 125 =$

6. $125 \times 96 =$

7. $12 \times 33 =$

8. $16 \times 9 =$

Math Masters

9. $9 \times 2.4 =$

10. $120 \times 4.5 =$

11. $160 \times 1.25 =$

12. $600 \div 12.5 =$

(See answers on page 196)

NUMBER CURIOSITY

$$6^2 - 5^2 = 11$$
$$56^2 - 45^2 = 1,111$$
$$556^2 - 445^2 = 111,111$$
$$5,556^2 - 4,445^2 = 11,111,111$$
etc.

DAY 8:

MULTIPLICATION AND DIVISION VI

DAY 8

TECHNIQUE 29:
Dividing by 15

BASIC APPLICATION

You begin a trip with a full gas tank, and travel 270 miles before stopping at a gas station. It takes 15 gallons to refill the tank. How many miles per gallon has your car experienced?

Calculation: $270 \div 15 = ?$

The Old-Fashioned Way

```
      18
15)270
      15
     ---
     120
     120
     ---
       0
```

The Excell-erated Way

Dividing by 15 is the same as multiplying by 2, and dividing by 30. Stated differently, it's identical with multiplying by ⅔ and then checking for decimal insertion. The best way to take ⅔ is to first take ⅓, and then double it. Using the above example, we simply take ⅔ of 27 (forget the zero), and we've got the answer! Let's take it a little more slowly, below.

133

To divide 270 by 15, answer the following questions

A. How does the calculation look, omitting the zero?   Answer: $27 \div 15$

B. What is $27 \times \frac{2}{3}$?   Answer: 18 (tentative solution)

[NOTE: Did you remember to first take $\frac{1}{3}$ of 27, then double it?]

Now let's test the tentative solution to see if it appears to be the solution to $270 \div 15$. How about working backwards, and asking yourself if $18 \times 15 = 270$. Using Technique #21 (Multiplying by 15), you can see that the above multiplication is valid, and 18 is the correct solution.

Let's Try One More Basic Calculation: $330 \div 15$

To divide 330 by 15, answer the following questions

A. How does the calculation look, omitting the zero?   Answer: $33 \div 15$

B. What is $33 \times \frac{2}{3}$?   Answer: 22 (tentative solution)

Performing the same type of test on the tentative solution as was performed above, you'll see that 22 is the correct solution, as well.

Math-Master Application

This technique will also work when dividing by 0.15, 1.5, 150, and so on. Let's assume you have just made a bulk purchase at your local hardware superstore of 150 large plastic bags for $18. What was the cost per bag?

Calculation: $18 \div 150 = ?$

To divide $18 by 150, answer the following questions

A. What does the calculation look like, omitting the dollar sign and zero?   Answer: $18 \div 15$

B. What is $18 \times \frac{2}{3}$?   Answer: $12 (the tentative solution)

Let's test the tentative solution. We know, simply by inspection, that the solution to \$18 ÷ 150 must be under \$1. Let's try \$.12 (12 cents) and work backwards. If we take 150 × \$.12, we do obtain \$18. Therefore, \$.12 is the correct solution.

Food for Thought

The real key to applying this technique is multiplying by ⅔. You can either first multiply by 2 and then divide by 3, or first divide by 3 and then multiply by 2 (whichever is easier for you). Remember that you can instead use Technique #18 (Dividing by 1.5, 2.5, etc.) to divide by 15. Choose whichever method is easier under the circumstances.

NOW IT'S YOUR TURN

Basic Exercises

1. $810 \div 15 =$
2. $180 \div 15 =$
3. $48 \div 15 =$
4. $720 \div 15 =$
5. $93 \div 15 =$
6. $120 \div 15 =$
7. $90 \div 15 =$
8. $84 \div 15 =$
9. $600 \div 15 =$
10. $360 \div 15 =$
11. $990 \div 15 =$
12. $69 \div 15 =$
13. $420 \div 15 =$

Math Masters

14. $5,400 \div 150 =$
15. $24 \div 1.5 =$
16. $3,900 \div 15 =$
17. $2,100 \div 150 =$
18. $6.3 \div 1.5 =$
19. $960 \div 150 =$
20. $66,000 \div 15 =$

(See answers on page 197)

DAY 8

TECHNIQUE 30:
Squaring Any Two-Digit Number Beginning in 5

BASIC APPLICATION

Your health insurance premiums are $52 per week. How much, therefore, do you pay in premiums per year?

Calculation: $52 × 52 = ?

The Old-Fashioned Way

$$\begin{array}{r} \overset{\scriptstyle 1}{\$52} \\ \times\, 52 \\ \hline 104 \\ 2\ 60 \\ \hline \$2,704 \end{array}$$

The Excell-erated Way

This technique will work when squaring numbers in the 50's. You begin the calculation by adding 25 to the ones digit. This will constitute the left portion of the answer. Then you simply square the ones digit for the right portion of the answer. One detail—when squaring 1, 2, or 3, write "01," "04," and "09," respectively. Let's apply this technique to our insurance premium problem.

To multiply 52 by 52, answer the following questions

A. What is 25 + 2?   Answer: 27 (left portion of solution)

B. What is 2 × 2?   Answer: 04 (right portion of solution)

C. What are the above answers when combined?   Answer: 2,704 (or $2,704)

Let's Try One More Basic Calculation:

To multiply 57 by 57, answer the following questions

A. What is 25 + 7?   Answer: 32 (left portion of solution)

B. What is 7 × 7?   Answer: 49 (right portion of solution)

C. What are the above answers when combined?   Answer: 3,249

Math-Master Application

Even though a calculation such as 5.6 × 560 does not constitute a square, this technique will still apply. Let's assume you've written a book which is an instant hit. Your first royalty check reflects the 56,000 copies that have been sold. If you earn $.56 (56 cents) in royalties per book, what will your royalty check total?

Calculation: 56,000 × $.56 = ?

To multiply 56,000 by $.56, answer the following questions

A. What does the calculation look like, omitting the dollar sign, decimal point, and zeroes?   Answer: 56 × 56

B. What is 25 + 6?   Answer: 31 (left portion of tent. solution)

C. What is 6 × 6?   Answer: 36 (right portion of tent. solution)

D. What are the above answers when combined?   Answer: 3,136 (or $3,136) (tentative solution)

Let's reexamine the original calculation to test the tentative solution. We know that $.56 is slightly more than half of a dollar. Therefore, $56,000 \times \$.56$ must be slightly more than half of $56,000, or $28,000. If we tack a zero onto the tentative solution, we obtain the correct solution, $31,360.

Food for Thought

When applying this technique, you must remember to add 25 to the ones digit to obtain the left portion of the solution. The number 25 is easy to remember because it is equal to 5^2, and 5 is the tens digit of the number to be squared. Try squaring 51, and then use Technique #19 (Squaring Any Number Ending in 1) to check your answer!

NOW IT'S YOUR TURN

Basic Exercises

1. $53^2 =$
2. $58^2 =$
3. $51^2 =$
4. $59 \times 59 =$
5. $56 \times 56 =$
6. $54 \times 54 =$
7. $55^2 =$
8. $57^2 =$
9. $52^2 =$
10. $58 \times 58 =$
11. $51 \times 51 =$
12. $53 \times 53 =$
13. $56^2 =$

Math Masters

14. $5.4 \times 5.4 =$

15. $5.2 \times 520 =$

16. $510^2 =$

17. $0.59 \times 59 =$

18. $550 \times 5.5 =$

19. $57 \times 570 =$

20. $5.3^2 =$

(See answers on page 197)

NUMBER CURIOSITY

$$(1)^2 = \quad 1 = 1^3$$
$$(1 + 2)^2 = \quad 9 = 1^3 + 2^3$$
$$(1 + 2 + 3)^2 = \quad 36 = 1^3 + 2^3 + 3^3$$
$$(1 + 2 + 3 + 4)^2 = 100 = 1^3 + 2^3 + 3^3 + 4^3$$
$$(1 + 2 + 3 + 4 + 5)^2 = 225 = 1^3 + 2^3 + 3^3 + 4^3 + 5^3$$
$$\text{etc.}$$

DAY 8

TECHNIQUE 31:
Multiplying by Regrouping

BASIC APPLICATION

The school your children attend owns and operates six buses, all of which can accommodate 43 children. How many children in all can be transported by bus?

Calculation: 43 × 6 = ?

The Old-Fashioned Way

```
  |
  43
× 6
 258
```

The Excell-erated Way

This technique is very powerful, and requires some creativity in applying. It essentially requires that you break down the calculation into two or three separate calculations. In our bus example, the calculation can be regrouped as $(40 \times 6) + (3 \times 6)$. When using this trick, it's especially important to apply the concept of "place value." In our bus calculation, the "4" of "43" actually represents "40," and it should be treated as such. Let's look a little more closely at our example.

141

To multiply 43 by 6, answer the following questions

A. What's another way to   Answer: $(40 + 3) \times 6$, or
    express 43×6? $(40 \times 6) + (3 \times 6)$
B. What is 40×6?   Answer: 240
C. What is 3×6?   Answer: 18
D. What is $240 + 18$?   Answer: 258 (the solution to
                                 43×6)

Let's Try One More Basic Calculation:

To multiply 25 by 26, answer the following questions

(HINT: This requires imagination. For example, you can see that it's very close to squaring the number 25.)

A. What's another way to   Answer: $25 \times (25 + 1)$, or
    express 25×26? $(25 \times 25) + (25 \times 1)$
B. What is 25×25?   Answer: 625 (per Technique
                                 #14)
C. What is 25×1?   Answer: 25
D. What is $625 + 25$?   Answer: 650 (the solution to
                                 25×26)

Math-Master Application

In measuring your house prior to remodeling, you note that it is exactly 21 yards wide. How many inches does that equal?

Calculation: $36 \times 21 = ?$

To multiply 36 by 21, answer the following questions

A. What's another way to   Answer: $36 \times (20 + 1)$, or
    express 36×21? $(36 \times 20) + (36 \times 1)$
B. What is 36×20?   Answer: 720
C. What is 36×1?   Answer: 36
D. What is $720 + 36$?   Answer: 756 (the solution to
                                 36×21)

Food for Thought

In the calculation immediately above, you could have regrouped the calculation as $(30 + 6) \times 21$, which equals $(30 \times 21) + (6 \times 21)$, which equals 756. However, it's obviously more cumbersome to approach the problem this way. It's best to re-express the number with the smaller ones digit (1 instead of 6), because the resulting addition is far more manageable.

NOW IT'S YOUR TURN

Basic Exercises

1. $32 \times 7 =$
2. $83 \times 5 =$
3. $22 \times 9 =$
4. $4 \times 47 =$
5. $8 \times 62 =$
6. $6 \times 54 =$
7. $76 \times 3 =$
8. $38 \times 5 =$
9. $44 \times 7 =$
10. $6 \times 93 =$
11. $9 \times 52 =$
12. $4 \times 87 =$
13. $36 \times 8 =$

Math Masters

14. $16 \times 15 =$
15. $36 \times 22 =$

16. $18 \times 31 =$

17. $45 \times 46 =$

18. $21 \times 33 =$

19. $22 \times 45 =$

20. $30 \times 105 =$

(See answers on page 197)

DAY 8

TECHNIQUE 32:
Multiplying by Augmenting

BASIC APPLICATION

You have exactly 19 weeks to complete a project. How many days, therefore, do you have before the deadline arrives?

Calculation: $19 \times 7 = ?$

The Old-Fashioned Way

$$\begin{array}{r} \overset{6}{19} \\ \times\ 7 \\ \hline 133 \end{array}$$

The Excell-erated Way

This technique is similar to Technique #31, and just as powerful. In this scenario, however, one of the numbers to multiply is just under a power of ten, such as 19 or 48. In our project problem above, we momentarily pretend that the number 19 is really 20. We then restate 19×7 as $(20 - 1) \times 7$, and go from there. Let's see how this indispensable method works in step-by-step fashion.

145

To multiply 19 by 7, answer the following questions

A. What's another way to   Answer: $(20 - 1) \times 7$, or
    express 19×7? $(20 \times 7) - (1 \times 7)$
B. What is 20×7?   Answer: 140
C. What is 1×7?   Answer: 7
D. What is $140 - 7$?   Answer: 133 (the solution to
                                  19×7)

Let's Try One More Basic Calculation: 6×28

To multiply 6 by 28, answer the following questions

A. What's another way to   Answer: $6 \times (30 - 2)$, or
    express 6×28? $(6 \times 30) - (6 \times 2)$
B. What is 6×30?   Answer: 180
C. What is 6×2?   Answer: 12
D. What is $180 - 12$?   Answer: 168 (the solution to
                                  6×28)

Math-Master Application

You empty your piggy bank and find that you only have quarters, and you count 39 in all. How much money do you have?

Calculation: $39 \times \$.25 = ?$

To multiply 39 by $.25, answer the following questions

A. Omitting the dollar sign and   Answer: $(40 - 1) \times 25$, or
    decimal, what's another way $(40 \times 25) - (1 \times 25)$
    to express $39 \times \$.25$?
B. What is 40×25?   Answer: 1,000
C. What is 1×25?   Answer: 25
D. What is $1,000 - 25$?   Answer: 975 (tentative solution)

Let's place a dollar sign in front of the tentative solution, and determine where to place the decimal point. We know that 4 quarters equal a dollar, and 40 quarters equal 10 dollars. So $39 \times \$.25$ must equal one quarter under $10, or $9.75.

Food for Thought

Though similar to Technique #31, this technique requires a bit more concentration because you have to subtract after performing and storing two other calculations. With practice, however, you'll find that this is one of the best tricks in the book. The theory behind both Techniques #31 and #32, of course, is that it is easier to multiply by a power of ten than by a number slightly above or below that power of ten.

NOW IT'S YOUR TURN

Basic Exercises

1. $29 \times 5 =$
2. $58 \times 7 =$
3. $39 \times 6 =$
4. $8 \times 48 =$
5. $3 \times 89 =$
6. $4 \times 68 =$
7. $19 \times 7 =$
8. $99 \times 6 =$
9. $78 \times 3 =$
10. $5 \times 69 =$
11. $3 \times 88 =$
12. $9 \times 49 =$
13. $28 \times 4 =$

Math Masters

14. $59 \times 30 =$
15. $98 \times 15 =$
16. $19 \times 70 =$

17. $25 \times 29 =$

18. $45 \times 18 =$

19. $15 \times 79 =$

20. $99 \times 12 =$

(See answers on page 197)

DAY 8 REVIEW QUIZ

DIRECTIONS: Work each exercise as quickly as possible, using the techniques just learned.

Basic Exercises

1. $360 \div 15 =$ 5. $38 \times 3 =$

2. $54^2 =$ 6. $54 \times 8 =$

3. $42 \times 7 =$ 7. $58 \times 58 =$

4. $6 \times 49 =$ 8. $480 \div 15 =$

Math Masters

9. $930 \div 150 =$ 11. $22 \times 45 =$

10. $5.7 \times 5.7 =$ 12. $19 \times 30 =$

(See answers on page 197)

DAY 9:

ESTIMATION I

DAY 9

TECHNIQUE 33:
Estimating Multiplication by 33 or 34

BASIC APPLICATION

Your copy machine will make 33 uninterrupted copies per minute. Approximately how many copies will it make in 15 minutes?

Calculation: $15 \times 33 \approx ?$

The Old-Fashioned Way

Unfortunately, estimation techniques are rarely taught in school, except, perhaps, for "rounding."

The Excell-erated Way

If we were multiplying by 33⅓, we would simply divide by 3 and then multiply by 100 to obtain an exact answer. Well, the numbers 33 and 34 are close enough to 33⅓ to justify using the same technique to obtain a very good estimate. Let's try our first estimation technique on our "copy" example above. (The " $\approx$ " symbol means "approximately equals.")

To estimate 15 × 33, answer the following questions

A.   What is 15 ÷ 3?   Answer: 5
B.   What is 5 × 100?   Answer: 500 (Our estimate. The exact
                               answer is 495, or 1% lower)

Let's Try One More Basic Calculation: Estimate 34 × 21

To estimate 34 × 21, answer the following questions

A.   What is 21 ÷ 3?   Answer: 7
B.   What is 7 × 100?   Answer: 700 (Our estimate. The exact
                               answer it 714, or 2% higher)

Math-Master Application

A professional baseball team wishes to stock up on baseballs by purchasing six dozen @ $3.40 apiece. What will be the approximate total cost?

Calculation: 72 × $3.40 ≈ ?

To estimate 72 × $3.40, answer the following questions

A.   What does the calculation   Answer: 72 × 34
      look like, omitting the dollar
      sign, decimal point, and
      zero?
B.   What is 72 ÷ 3?   Answer: 24, or $24 (tentative
                                       estimate)

Let's test the tentative estimate. Looking at the original calculation, 72 × $3.40, we can see that the answer is somewhere in the $200's. Therefore, we'll need to tack a zero onto the tentative estimate to produce our final estimate, $240. The exact answer, incidentally, is $244.80, or 2% higher.

Food for Thought

This was the first of eight estimation techniques presented in this book. The primary criteria used to determine which estimation techniques, out of the many available, would be presented were (1) frequency of use, and (2) accuracy to within two percent. The key to Technique #33 is to divide by 3 when estimating multiplication by 33 or 34.

NOW IT'S YOUR TURN

Basic Exercises

1. $18 \times 33 \approx$
2. $27 \times 33 \approx$
3. $15 \times 34 \approx$
4. $48 \times 34 \approx$
5. $33 \times 12 \approx$
6. $33 \times 21 \approx$
7. $34 \times 66 \approx$
8. $34 \times 57 \approx$
9. $72 \times 33 \approx$
10. $45 \times 33 \approx$
11. $24 \times 34 \approx$
12. $99 \times 34 \approx$
13. $33 \times 33 \approx$

Math Masters

14. $33 \times 93 \approx$
15. $34 \times 84 \approx$
16. $3.4 \times 75 \approx$
17. $8.7 \times 3.3 \approx$
18. $3.9 \times 330 \approx$
19. $3.4 \times 81 \approx$
20. $34 \times 5.4 \approx$

(See answers on page 198)

DAY 9

TECHNIQUE 34:
Estimating Division by 33 or 34

BASIC APPLICATION

Last week, you worked 34 hours and earned gross wages of $220. What was your approximate wage per hour?

Calculation: $220 ÷ 34 ≈ ?

The Old-Fashioned Way

None, for estimating

The Excell-erated Way

If we were dividing by 33⅓, we would simply multiply by 3 and then divide by 100 to obtain an exact answer. As expressed in Technique #33, the numbers 33 and 34 are very close to 33⅓. We can, therefore, use its techniques to obtain a good estimate. Let's try this technique on our "wage" example above.

To estimate $220 ÷ 34, answer the following questions

A.   What is $220 × 3?   Answer: $660
B.   What is $660 ÷ 100?   Answer: $6.60 (Our estimate. The exact answer is about $6.47, or 2% lower)

(NOTE: You could instead take $22 × 3, but then divide by 10 to obtain the estimate.)

Let's Try One More Basic Calculation: Estimate 170 ÷ 33

To estimate 170 ÷ 33, answer the following questions

A. What is 170 × 3?   Answer: 510

B. What is 510 ÷ 100?   Answer: 5.10, or 5.1 (Our estimate.
                              The exact answer is about 5.15, or 1%
                              higher.)

(NOTE: You could instead take 17 × 3, but then divide by 10 to obtain the estimate.)

Math-Master Application

Every time you jog around the block, you cover 330 yards. To cover 7,000 yards, you'll need to circle the block approximately how many times?

Calculation: 7,000 ÷ 330 ≈ ?

To estimate 7,000 ÷ 330, answer the following questions

A. What does the calculation   Answer: 7 ÷ 33
    look like, omitting all zeroes?

B. What is 7 × 3?   Answer: 21 (tentative estimate)

Let's test the tentative estimate by going back to the original calculation. If we lop off a zero from the dividend and divisor, we get 700 ÷ 33. Working backwards, we ask ourselves if 21 × 33 appears to approximately equal 700, or if we need to alter the tentative estimate by tacking on zeroes or inserting a decimal point. By inspection, we can see that 21 is, in fact, the final estimate. The exact answer, incidentally, is about 21.21, or 1% higher.

Food for Thought

The key to Technique #34 is to multiply by 3 when estimating division by 33 or 34. In the unusual event that you are multiplying or dividing by 33⅓, you can obtain an *exact* answer by applying Techniques #33 and #34, respectively. These two techniques will work, incidentally, with numbers between 330 and 340 (as 336), between 3,300 and 3,400 (as 3,342), and so forth. Just proceed in the same fashion, and be sure to always test the tentative estimate.

NOW IT'S YOUR TURN

Basic Exercises

1. $260 \div 33 \approx$
2. $600 \div 33 \approx$
3. $210 \div 34 \approx$
4. $700 \div 34 \approx$
5. $140 \div 33 \approx$
6. $800 \div 33 \approx$
7. $230 \div 34 \approx$
8. $500 \div 34 \approx$
9. $110 \div 33 \approx$
10. $400 \div 33 \approx$
11. $160 \div 34 \approx$
12. $250 \div 34 \approx$
13. $750 \div 33 \approx$

Math Masters

14. $280 \div 3.3 \approx$
15. $1,200 \div 34 \approx$
16. $200 \div 3.4 \approx$
17. $520 \div 330 \approx$
18. $3,100 \div 330 \approx$

19. $13 \div 3.4 \approx$

20. $1.8 \div 0.34 \approx$

(See answers on page 198)

NUMBER CURIOSITY

$987,654,321 \times 2 = 1,975,308,642$

$987,654,321 \times 4 = 3,950,617,284$

$987,654,321 \times 5 = 4,938,271,605$

$987,654,321 \times 7 = 6,913,580,247$

$987,654,321 \times 8 = 7,901,234,568$

(Each product contains the digits 0–9 exactly once)

DAY 9

TECHNIQUE 35:
Estimating Multiplication by 49 or 51

BASIC APPLICATION

You retire exactly 7 weeks (or 49 days) from this moment. Approximately how many hours is that from now?

Calculation: $24 \times 49 \approx$?

The Old-Fashioned Way

None, for estimating

The Excell-erated Way

If we were multiplying by 50, we would simply divide by 2 and then multiply by 100. The numbers 49 and 51 are close enough to 50 to justify using the same technique to estimate. Let's try our "retirement" example above to apply this estimation technique.

To estimate 24×49, answer the following questions

A. What is $24 \div 2$?   Answer: 12
B. What is 12×100?   Answer: 1,200 (Our estimate. The exact answer is 1,176, or 2% lower.)

Let's Try One More Basic Calculation: Estimate 51 × 76

To estimate 51 × 76, answer the following questions

A. What is 76 ÷ 2?   Answer: 38
B. What is 38 × 100?   Answer: 3,800 (Our estimate. The
                         exact answer is 3,876, or 2% higher.)

Math-Master Application

A theatre sells 380 tickets during a given day. If each ticket sells for $5.10, what are approximate total sales?

Calculation: 380 × $5.10 ≈ ?

To estimate 380 × $5.10, answer the following questions

A. What does the calculation   Answer: 38 × 51
    look like, omitting the zeroes,
    dollar sign, and decimal
    point?
B. What is 38 ÷ 2?   Answer: 19, or $19 (tentative
                                    estimate)

Let's test the tentative estimate. If we round the original calculation to 400 × $5, we know that the answer is around $2,000. Therefore, if we tack two zeroes onto the tentative estimate, we obtain the final estimate, $1,900. The exact answer, incidentally, is $1,938, or 2% higher.

Food for Thought

The key to Technique #35 is to divide by 2 when estimating multiplication by 49 or 51. Remember that whatever technique you are applying, always check your intended answer against the original calculation to see if it "looks right," much like you would check the spelling of a word by seeing if it looks right. Also remember that the tests of the tentative answers and estimates are merely suggestions and not set in stone.

NOW IT'S YOUR TURN

Basic Exercises

1. $34 \times 49 \approx$
2. $18 \times 49 \approx$
3. $26 \times 51 \approx$
4. $66 \times 51 \approx$
5. $49 \times 72 \approx$
6. $49 \times 58 \approx$
7. $51 \times 13 \approx$
8. $51 \times 87 \approx$
9. $29 \times 49 \approx$
10. $41 \times 49 \approx$
11. $84 \times 51 \approx$
12. $99 \times 51 \approx$
13. $49 \times 33 \approx$

Math Masters

14. $4.9 \times 38 \approx$
15. $51 \times 1.4 \approx$
16. $510 \times 0.96 \approx$
17. $7.6 \times 49 \approx$
18. $610 \times 4.9 \approx$
19. $2.4 \times 51 \approx$
20. $0.92 \times 510 \approx$

(See answers on page 199)

DAY 9

TECHNIQUE 36:
Estimating Division by 49 or 51

BASIC APPLICATION

You've just retired, and your 51 coworkers have chipped in a total of $620 to pay for a retirement party and gift. Approximately what was the average donation per person?

Calculation: $620 ÷ 51 ≈ ?

The Old-Fashioned Way

None, for estimating

The Excell-erated Way

If we were dividing by 50, we would simply multiply by 2 and then divide by 100. Because the numbers 49 and 51 are very close to 50, we can use the above technique to estimate. Let's use this technique to figure out how generous your coworkers were.

To estimate $620 ÷ 51, answer the following questions

A.   What is $620 × 2?   Answer: $1,240 (or simply double $62,
                                then tack on a zero)
B.   What is $1,240 ÷ 100?   Answer: $12.40 (Our estimate. The
                                actual answer is about $12.16, or 2%
                                lower.)

Let's Try One More Basic Calculation: Estimate 260 ÷ 49

To estimate 260 ÷ 49, answer the following questions

A.   What is 260 × 2?   Answer: 520
B.   What is 520 ÷ 100?   Answer: 5.20, or simply 5.2 (Our
                                estimate. The actual answer is about
                                5.3, or 2% higher.)

Math-Master Application

You've discovered the sale of the century. At your local music store, you
can purchase an unlimited number of CD's at $4.90 apiece. If you've
decided to use all of the $140 you received at your retirement party,
approximately how many CD's can you buy?

Calculation: $140 ÷ $4.90 ≈ ?

To estimate $140 ÷ $4.90, answer the following questions

A.   What does the calculation look like,   Answer: 14 ÷ 49
     omitting the dollar sign, decimal point,
     and zeroes?
B.   What is 14 × 2?
                                               Answer: 28
                                               (tentative estimate)

Let's test the tentative estimate. Working backwards, we ask ourselves if
28 × $4.90 (or 28 × about $5.00) equals $140. It does, so the actual
estimate is 28. The actual answer, incidentally, is about 28.6, or 2%
higher.

Food for Thought

The key to Technique #36 is to multiply by 2 when estimating division by
49 or 51. Techniques #35 and #36 will work, incidentally, with numbers

between 490 and 510 (such as 495), or between 4,900 and 5,100 (such as 5,032), and so forth. Just proceed in the same fashion, and be sure to always test the tentative estimate.

NOW IT'S YOUR TURN

Basic Exercises

1. $330 \div 49 \approx$
2. $95 \div 49 \approx$
3. $640 \div 51 \approx$
4. $410 \div 51 \approx$
5. $160 \div 49 \approx$
6. $720 \div 49 \approx$
7. $270 \div 51 \approx$
8. $690 \div 51 \approx$
9. $380 \div 49 \approx$
10. $840 \div 49 \approx$
11. $190 \div 51 \approx$
12. $87 \div 51 \approx$
13. $910 \div 49 \approx$

Math Masters

14. $56 \div 4.9 \approx$
15. $1,400 \div 510 \approx$
16. $2,200 \div 51 \approx$
17. $4.4 \div 0.49 \approx$
18. $76 \div 4.9 \approx$
19. $2.4 \div 0.51 \approx$
20. $3,500 \div 510 \approx$

(See answers on page 199)

JUST FOR FUN

How many people would you have to put in a room, whereby the probability is about 50% that at least two people share the same birthday?

ANSWER: 23 people. (The strategy, oddly enough, is to determine the probability that no two people share the same birthday. We therefore take (365/365) × (364/365) × (363/365), and so on, until we produce a product of approximately 0.5 (50% probability). Each multiplier here represents one person. When we carry the calculation out to (365/365) × ... × (343/365), representing 23 persons, we produce a likelihood of 49.2697% (or 0.492697) that no two people share the same birthday. Therefore, there is slightly more than a 50% chance that two *do* share a birthday.)

DAY 9 REVIEW QUIZ

DIRECTIONS: Work each estimation exercise as quickly as possible, using the techniques just learned.

Basic Exercises

1. $24 \times 33 \approx$

2. $140 \div 34 \approx$

3. $72 \times 51 \approx$

4. $270 \div 49 \approx$

5. $380 \div 51 \approx$

6. $49 \times 66 \approx$

7. $250 \div 33 \approx$

8. $34 \times 48 \approx$

Math Masters

9. $330 \times 0.84 \approx$

10. $120 \div 3.4 \approx$

11. $490 \times 1.8 \approx$

12. $7,600 \div 510 \approx$

(See answers on page 200)

FISH STORY?

I remember watching (a long time ago) a TV game show called "I've Got a Secret," in which celebrity panelists had to guess the unusual secret of non-celebrity guests. On one show, four women claimed to have each been dealt a perfect bridge hand on the same deal (i.e., all 13 cards of the same suit).

Hold on there! The probability of just *one* player being dealt such a hand is 1 in over 158 BILLION. I suppose anything's possible, but I also believe that Barnum was right—there *is* a sucker born every minute!

FOOTNOTE: It's impossible to deal exactly three perfect bridge hands on one deal. Think about it.

DAY 10:

ESTIMATION II

DAY 10

TECHNIQUE 37:
Estimating Multiplication by 66 or 67

BASIC APPLICATION

You're about to take a long trip by automobile. You figure you average about 66 mph on the open highway. If you expect to spend a total of 18 hours of driving time over the next three days, approximately how many miles will you cover?

Calculation: $18 \times 66 \approx ?$

The Old-Fashioned Way

None, for estimating

The Excell-erated Way

If we were multiplying by 66⅔, we would simply multiply by ⅔ and then multiply by 100 to obtain an exact answer. Because the numbers 66 and 67 are very close to 66⅔, we can apply the same technique to obtain a good estimate. To multiply by ⅔, it's easiest to take ⅓, and then double it. For example, ⅔ of 21 equals ⅓ of 21, or 7, doubled, or 14.

169

To estimate 18 × 66, answer the following questions

A.   What is ⅔ of 18?   Answer: 12 (Did you take ⅓ of 18 and
                                                   then double it?)

B.   What is 12 × 100?   Answer: 1,200 (Our estimate. The exact
                                                   answer is 1,188, or 1% lower.)

Let's Try One More Basic Calculation: Estimate 67 × 75

To estimate 75 × 67, answer the following questions

A.   What is ⅔ of 75?   Answer: 50
B.   What is 50 × 100?   Answer: 5,000 (Our estimate. The exact
                                                   answer is 5,025, or ½ of 1% higher.)

Math-Master Application

Finally, the $240 coat you've been eyeing is marked down 33%. What is
the approximate sale price? (Math Tip: To obtain the sale price of an item
marked down 33%, simply multiply the regular price by 67%.)

Calculation: $240 × 67% ≈
                    (or $240 × 0.67)

To estimate $240 × 0.67, answer the following questions

A.   What does the calculation   Answer: 24 × 67
       look like, omitting the dollar
       sign, decimal point, and
       zeroes?

B.   What is ⅔ of 24?   Answer: 16, or $16 (tentative
                                                   estimate)

By inspection of the original calculation, we can determine that one zero
must be tacked on to the tentative estimate, to produce our final estimate, $160. The exact answer, incidentally, is $160.80, or ½ of 1% higher.

Food for Thought

Technique #37 is very similar to Technique #23 (Multiplying by 75), in
that they both use fractions to solve problems expressed in decimal form.
To multiply by 75, we begin by multiplying by ¾. Similarly, to multiply by

66 or 67, we begin by multiplying by ⅔. Remember, though, that the former calculation will produce an exact answer, while the latter will produce an estimate.

NOW IT'S YOUR TURN

Basic Exercises

1. $42 \times 66 \approx$
2. $24 \times 66 \approx$
3. $36 \times 67 \approx$
4. $12 \times 67 \approx$
5. $66 \times 45 \approx$
6. $66 \times 27 \approx$
7. $67 \times 21 \approx$
8. $67 \times 54 \approx$
9. $60 \times 66 \approx$
10. $33 \times 66 \approx$
11. $72 \times 67 \approx$
12. $30 \times 67 \approx$
13. $66 \times 15 \approx$

Math Masters

14. $66 \times 87 \approx$
15. $67 \times 78 \approx$
16. $6.7 \times 9.9 \approx$
17. $480 \times 0.66 \approx$
18. $3.9 \times 66 \approx$
19. $0.96 \times 670 \approx$
20. $8.1 \times 6.7 \approx$

(See answers on page 200)

DAY 10

TECHNIQUE 38:
Estimating Division by 66 or 67

BASIC APPLICATION

You've just purchased the latest best-selling book, an 800-page political exposé. For a book such as this, you can read about 66 pages per hour. About how long will it take you to read the entire book?

Calculation: $800 \div 66 \approx ?$

The Old-Fashioned Way   The Excell-erated Way

None, for estimating

If we were dividing by 66⅔, we would simply multiply by 1½ (or 1.5) and then divide by 100 to obtain an exact answer. Because the numbers 66 and 67 are so close to 66⅔, we can use the same technique to obtain a good estimate. At this point, it would be advisable to review Technique #21 (Multiplying by 15), because the same strategy is used to multiply by 1.5. Let's now see how long it's going to take you to read that new book of yours.

To estimate 800 ÷ 66, answer the following questions

(NOTE: In this situation, it's easier to first divide by 100, and then multiply by 1.5.)

A. What is 800 ÷ 100? Answer: 8
B. What is 8 × 1.5? Answer: 12 (Our estimate. The actual answer is about 12.1, or 1% higher.)

Let's Try One More Basic Calculation: Estimate 1,400 ÷ 67

To estimate 1,400 ÷ 67, answer the following questions

(Here, too, it's best to first divide by 100, and then multiply by 1.5.)

A. What is 1,400 ÷ 100? Answer: 14
B. What is 14 × 1.5? Answer: 21 (Our estimate. The actual answer is about 20.9, or ½ of 1% lower.)

Math-Master Application

It's the holiday season, and you've got 500 feet of ribbon to use in wrapping gifts. If the average gift requires 6.7 feet of ribbon, approximately how many gifts will you be able to wrap with your current supply of ribbon?

Calculation: 500 ÷ 6.7 ≈ ?

To estimate 500 ÷ 6.7, answer the following questions

A. What does the calculation Answer: 5 ÷ 67
   look like, omitting the zeroes
   and decimal point?
B. What is 5 × 1.5? Answer: 7.5 (tentative estimate)

Checking the tentative estimate by working in reverse, we ask ourselves if 7.5 × 6.7 appears to approximate 500. Obviously, it doesn't. However, 75 × 6.7 does appear to approximate 500, so 75 is our final estimate. The actual answer is about 74.6, or ½ of 1% lower than our estimate.

Food for Thought

To apply this technique, it's sometimes easier to first divide by 100, and then multiply by 1.5. Other times, it's easier to do the reverse. When performing a "Math-Master" calculation, whereby zeroes and decimal points are eliminated, you can simply multiply by 1.5, and immediately test the tentative estimate for reasonableness. Even though the number 1.5 has been used throughout our explanation, remember that you can use the equivalent mixed number, 1½, if you prefer.

NOW IT'S YOUR TURN

Basic Exercises

1. $320 \div 66 \approx$
2. $280 \div 66 \approx$
3. $160 \div 67 \approx$
4. $440 \div 67 \approx$
5. $520 \div 66 \approx$
6. $360 \div 66 \approx$
7. $1,800 \div 67 \approx$
8. $3,000 \div 67 \approx$
9. $1,200 \div 66 \approx$
10. $7,000 \div 66 \approx$
11. $460 \div 67 \approx$
12. $640 \div 67 \approx$
13. $580 \div 66 \approx$

Math Masters

14. $34 \div 6.6 \approx$
15. $8,000 \div 670 \approx$

16. $5,000 \div 6.7 \approx$

17. $26 \div 6.6 \approx$

18. $4.8 \div 0.66 \approx$

19. $5,600 \div 670 \approx$

20. $720 \div 6.7 \approx$

(See answers on page 201)

NUMBER CURIOSITY

$$123,456,789 \times 2 = 246,913,578$$
$$123,456,789 \times 4 = 493,827,156$$
$$123,456,789 \times 5 = 617,283,945$$
$$123,456,789 \times 7 = 864,197,523$$
$$123,456,789 \times 8 = 987,654,312$$
(Each product contains the digits 1–9 exactly once)

DAY 10

TECHNIQUE 39:
Estimating Division by 9

BASIC APPLICATION

On your last golf outing, you shot 43 for the first 9 holes. What was your approximate stroke average per hole?

Calculation: $43 \div 9 \approx ?$

The Old-Fashioned Way

None, for estimating

The Excell-erated Way

Because the numbers 9 and 11 multiply out to close to 100, you can estimate division by 9 by multiplying by 11, and estimate division by 11 by multiplying by 9. In either case, you would then have to divide by 100. Before applying Technique #39, you should briefly review Technique #13 (Multiplying by 11). Let's test this estimation technique on our golf score.

To estimate 43 ÷ 9, answer the following questions

A. What is 43 × 11?   Answer: 473 (Did you use the "11 Trick"?)

B. What is 473 ÷ 100?   Answer: 4.73 (Our estimate. The actual answer is about 4.78, which is 1% higher.)

176

Let's Try One More Basic Calculation: Estimate 87 ÷ 9

To estimate 87 ÷ 9, answer the following questions

A. What is 87 × 11?   Answer: 957
B. What is 957 ÷ 100?   Answer: 9.57 (Our estimate. The actual
                         answer is about 9.67, which is 1% higher.)

Math-Master Application

The business you own has just experienced a good year, and you would
like to express your appreciation to your 90 employees by distributing a
$35,000 bonus equally. Approximately how much would each employee
receive?

Calculation: $35,000 ÷ 90≈?

To estimate $35,000 ÷ 90 answer the following questions

A. What does the calculation   Answer: 35 ÷ 9
    look like, omitting the dollar
    sign and zeroes?
B. What is 35 × 11?   Answer: 385, or $385 (tentative
                                   estimate)

Let's go back to the original calculation to test our tentative estimate.
Lopping off one zero from the dividend and divisor, we get $3,500 ÷ 9.
We know that $3,500 ÷ 10 equals $350, so $3,500 ÷ 9 must equal a bit
more than $350. Accordingly, the tentative estimate, $385, is also our
final estimate. The actual answer is $388.88 (rounded), or 1% higher.

Food for Thought

The theory behind Techniques #39 and #40 is that it is normally easier
to multiply than to divide. And because we have already covered techniques for multiplying by 9 and 11 (Techniques #25 and #13, respectively), you should be able to apply these last two techniques
comfortably. Remember that the lower the number, the greater the
likelihood that you will encounter a calculation involving that number.
Numbers like 9 and 11 crop up in calculations all the time.

NOW IT'S YOUR TURN

Basic Exercises

1. $320 \div 9 \approx$

2. $440 \div 9 \approx$

3. $25 \div 9 \approx$

4. $61 \div 9 \approx$

5. $570 \div 9 \approx$

6. $480 \div 9 \approx$

7. $64 \div 9 \approx$

8. $16 \div 9 \approx$

9. $400 \div 9 \approx$

10. $510 \div 9 \approx$

11. $85 \div 9 \approx$

12. $290 \div 9 \approx$

13. $150 \div 9 \approx$

Math Masters

14. $97 \div 9 \approx$

15. $1,100 \div 90 \approx$

16. $35 \div 0.9 \approx$

17. $760 \div 90 \approx$

18. $8,000 \div 900 \approx$

19. $1.4 \div 0.9 \approx$

20. $100 \div 90 \approx$

(See answers on page 201)

DAY 10

TECHNIQUE 40:
Estimating Division by 11

BASIC APPLICATION

You're at the supermarket, comparison shopping for a bag of potato chips. The economy size bag of chips costs 9 cents per ounce. However, the medium size, 11-ounce bag is on sale for 90 cents. You're wondering if the medium size bag is a better deal, and would like to know what its approximate cost per ounce is.

Calculation: $90 \div 11 \approx$?

The Old-Fashioned Way

None, for estimating

The Excell-erated Way

As explained in Technique #39, you can estimate division by 11 simply by multiplying by 9, and then dividing by 100. Remember that Technique #25 explains how to multiply by 9 quickly. So in our potato chip problem above, we take 90×9, and then divide by 100. As you'll see below, this technique really is easy to apply.

To estimate 90 ÷ 11, answer the following questions

A.   What is 90 × 9?   Answer: 810
B.   What is 810 ÷ 100?   Answer: 8.10, or simply 8.1 (Our
                                          estimate. The actual answer is about 8.2,
                                          or 1% higher.)

(NOTE: You could take 9 × 9, and then divide by 10, to obtain the estimate, 8.1. As it turns out, the medium size bag of chips turned out to be less expensive per ounce than the economy size bag.)

Let's Try One More Basic Calculation: 500 ÷ 11

To estimate 500 ÷ 11, answer the following questions

(Let's first divide by 100, and then multiply by 9.)

A.   What is 500 ÷ 100?   Answer: 5
B.   What is 5 × 9?   Answer: 45 (Our estimate. The actual
                                        answer is about 45.45, or 1% higher.)

Math-Master Application

You have just learned that your boss got a raise, bringing his weekly salary to $7,000. This represents a 10 percent increase for him. Approximately what was his weekly salary before the raise?

Calculation: $7,000 ÷ 1.1 ≈ ?
                  ($7,000 ÷ 110%)

To estimate $7,000 ÷ 1.1, answer the following questions

A.   What does the calculation look   Answer: 7 ÷ 11
     like, omitting the dollar sign,
     zeroes, and decimal point?
B.   What is 7 × 9?   Answer: 63, or $63 (tentative
                                                      estimate)

By inspection of the original calculation, the solution must be just a bit under $7,000. If we, therefore, tack on a couple of zeroes to the tentative estimate, we obtain our final estimate of $6,300. The actual answer, incidentally, is $6,363.64 (rounded), or 1% higher.

Food for Thought

To sum up this technique, whenever you need to estimate division by 11 (or 1.1, 110, etc.), simply multiply by 9, and test for reasonableness. There are other techniques of this type. For example, to divide by 14, you need only multiply by 7, test the tentative estimate, and your answer will be only 2% off. To divide by 17, you need only multiply by 6, and so forth.

NOW IT'S YOUR TURN

Basic Exercises

1. $400 \div 11 \approx$

2. $140 \div 11 \approx$

3. $600 \div 11 \approx$

4. $100 \div 11 \approx$

5. $160 \div 11 \approx$

6. $230 \div 11 \approx$

7. $360 \div 11 \approx$

8. $700 \div 11 \approx$

9. $800 \div 11 \approx$

10. $150 \div 11 \approx$

11. $900 \div 11 \approx$

12. $200 \div 11 \approx$

13. $450 \div 11 \approx$

Math Masters

14. $24 \div 1.1 \approx$
15. $3,000 \div 110 \approx$
16. $560 \div 11 \approx$
17. $130 \div 1.1 \approx$
18. $3,500 \div 110 \approx$
19. $7.5 \div 1.1 \approx$
20. $890 \div 11 \approx$

(See answers on page 202)

NUMBER CURIOSITY

$12,345,679 \times 2 = 24,691,358$ (contains digits 1–9, except 7)
$12,345,679 \times 4 = 49,382,716$ (contains digits 1–9, except 5)
$12,345,679 \times 5 = 61,728,395$ (contains digits 1–9, except 4)
$12,345,679 \times 7 = 86,419,753$ (contains digits 1–9, except 2)
$12,345,679 \times 8 = 98,765,432$ (contains digits 1–9, except 1)

(NOTE ALSO: In each case, multiplier plus digit missing equals 9)

DAY 10 REVIEW QUIZ

DIRECTIONS: Work each estimation exercise as quickly as possible, using the techniques just learned.

Basic Exercises

1. $72 \times 66 \approx$ **5.** $150 \div 11 \approx$

2. $700 \div 67 \approx$ **6.** $44 \div 9 \approx$

3. $250 \div 9 \approx$ **7.** $640 \div 66 \approx$

4. $3,000 \div 11 \approx$ **8.** $67 \times 24 \approx$

Math Masters

9. $6.6 \times 3.9 \approx$ **11.** $43 \div 0.9 \approx$

10. $2,600 \div 670 \approx$ **12.** $4,000 \div 110 \approx$

(See answers on page 202)

JUST FOR FUN

What is the number of possible spare combinations in bowling? For example, there are 10 possibilities if you leave only one pin standing. Assume any combination of pins could be left standing.

ANSWER: $2^{10} - 1$, or 1,023, possible spare combinations. Why 2^{10} *minus one*? When you knock down all the pins on the first ball, there is no spare possibility!

FINAL EXAM

DIRECTIONS: Work each estimation excercise as quickly as possible, using the techniques just learned.

Basic Exercises

1. $9 + 4 + 7 + 3 + 3 + 3 + 8 =$
2. $35 + 24 + 6 + 42 + 18 + 30 =$
3. $29 + 97 =$
4. $91 - 55 =$
5. $143 - 89 =$
6. $130 - 72 =$
7. $180 \div 4 =$
8. $5 \times 46 =$
9. $67 \times 11 =$
10. $95^2 =$
11. $900 \div 25 =$
12. $22 \div 5\frac{1}{2} =$
13. $36 \times 15 =$
14. $109 \times 104 =$
15. $88 \times 75 =$
16. $360 \div 75 =$
17. $9 \times 14 =$
18. $35 \times 12 =$
19. $72 \times 125 =$
20. $4,000 \div 125 =$
21. $420 \div 15 =$
22. $53^2 =$

23. $54 \times 6 =$

24. $7 \times 39 =$

25. $21 \times 33 \approx$

26. $500 \div 34 \approx$

27. $14 \times 49 \approx$

28. $160 \div 51 \approx$

29. $66 \times 27 \approx$

30. $1,200 \div 67 \approx$

31. $400 \div 9 \approx$

32. $140 \div 11 \approx$

Math Masters

33. $72 + 21 + 8 + 66 + 37 =$

34. $839 - 680 =$

35. $40 \times 5.1 =$

36. $3,900 \div 500 =$

37. $8\frac{1}{2} \times 800 =$

38. $1.9 \times 2.1 =$

39. $180 \times 1.5 =$

40. $122 \times 104 =$

41. $600 \div 7.5 =$

42. $6.5 \times 120 =$

43. $30 \div 1.25 =$

44. $520^2 =$

45. $21 \times 18 =$

46. $40 \div 3.3 \approx$

47. $1.8 \times 49 \approx$

48. $6.7 \times 21 \approx$

49. $43 \div 0.9 \approx$

50. $900 \div 110 \approx$

(See answers on page 203)

CONCLUSION

Congratulations! Your patience and persistence have paid off, as you have now completed the program. Now that you have learned the tricks of the trade, let me make some suggestions to enhance your mental-math wizardry.

First of all, glance through the 40 techniques, decide which ones you like the most, and use them. Remember, the use of only a dozen of these techniques will definitely make a difference. It's probably better to apply only the basics of, say, 20 techniques, than to apply both the basic and advanced (Math Master) concepts of only 10.

If you find yourself applying a technique in a slightly different way, or have discovered one not covered in this book, by all means use it (even if it only saves you a fraction of a second). Don't be afraid to experiment and to use your imagination when performing mental math. Whatever produces the correct answer the quickest is the technique you should use.

Be careful, however, to avoid techniques you've discovered or learned elsewhere that really don't save any time. I assure you I'm aware of a number of really fascinating techniques that, unfortunately, are so difficult to remember and apply that I never use them.

Most of all, practice these techniques every day, and be sure to keep your mind "mathematically fit." Whenever you see a couple of numbers, as on a license plate, billboard, or TV screen, manipulate those numbers in some way, even though you have no use for the answer. Remember, practice does make perfect.

In any case, I'd love to hear from you. Perhaps you have some input about this Excell-erated math program (favorable or unfavorable) that you would like to send my way. Or maybe I've left out your favorite rapid math technique, which you would like me to learn. Whatever the case may be, you can write to me at California Lutheran University, c/o School of Business, 60 West Olsen Road, Thousand Oaks, CA 91360. I will be sure to write back.

Also, I do hold workshops for schools, businesses, and organizations on the topic of rapid calculation. Please write to me at the above address for further information.

In closing, I hope you enjoyed the course as much as I enjoyed teaching it. Thank you, and good luck!

—ED JULIUS

ANSWERS TO PRACTICE EXERCISES AND EXAMS

Answers: Technique 1

Basic Exercises		Math Masters
1. 30	8. 43	14. 41
2. 33	9. 33	15. 44
3. 43	10. 36	16. 60
4. 34	11. 35	17. 45
5. 39	12. 43	18. 44
6. 37	13. 28	19. 49
7. 31		20. 45

Answers: Technique 2

Basic Exercises		Math Masters
1. 129	8. 144	14. 212
2. 134	9. 151	15. 246
3. 138	10. 155	16. 204
4. 142	11. 147	17. 236
5. 133	12. 169	18. 186
6. 131	13. 164	19. 223
7. 139		20. 275

Answers: Technique 3

Basic Exercises		Math Masters
1. 135	7. 120	13. 184
2. 107	8. 108	14. 222
3. 139	9. 131	15. 162
4. 176	10. 135	16. 173
5. 135	11. 149	17. 208
6. 129	12. 124	18. 190
		19. 158
		20. 253

Answers: Technique 4

Basic Exercises		Math Masters
1. 112	8. 75	14. 224
2. 123	9. 113	15. 172
3. 123	10. 131	16. 172
4. 126	11. 94	17. 195
5. 132	12. 123	18. 235
6. 114	13. 85	19. 184
7. 164		20. 242

Answers to Day 1 Review Quiz

Basic Exercises	Math Masters
1. 31	9. 45
2. 140	10. 251
3. 124	11. 225
4. 40	12. 225
5. 141	
6. 124	
7. 156	
8. 131	

Answers: Technique 5

Basic Exercises		Math Masters	
1. 45	7. 29	13. 37	18. 46
2. 22	8. 36	14. 75	19. 67
3. 37	9. 53	15. 59	20. 44
4. 18	10. 24	16. 65	
5. 37	11. 26	17. 48	
6. 48	12. 24		

Answers: Technique 6

Basic Exercises

1. 18
2. 49
3. 28
4. 72
5. 65
6. 56
7. 87
8. 54
9. 44
10. 37
11. 37
12. 68

Math Masters

13. 53
14. 26
15. 56
16. 27
17. 33
18. 48
19. 59
20. 37

Answers: Technique 7

Basic Exercises

1. 35
2. 37
3. 31
4. 33
5. 28
6. 36
7. 17
8. 34
9. 35
10. 37
11. 46
12. 35

Math Masters

13. 35
14. 71
15. 55
16. 38
17. 62
18. 23
19. 27
20. 69

Answers: Technique 8

Basic Exercises

1. 34
2. 61
3. 56
4. 77
5. 39
6. 23
7. 68
8. 51
9. 32
10. 57
11. 66
12. 49
13. 145

Math Masters

14. 34
15. 39
16. 66
17. 58
18. 85
19. 88
20. 48

Answers to Day 2 Review Quiz

Basic Exercises

1. 34
2. 27
3. 34
4. 47
5. 38
6. 47
7. 49
8. 56

Math Masters

9. 45
10. 56
11. 28
12. 56

Answers: Technique 9

Basic Exercises

1. 64
2. 108
3. 292
4. 208
5. 152
6. 256
7. 352
8. 76
9. 372
10. 124
11. 180
12. 264
13. 336

Math Masters

14. 1,120
15. 184
16. 38
17. 140
18. 284
19. 236
20. 396

Answers: Technique 10

Basic Exercises

1. 23
2. 14
3. 29
4. 55
5. 21
6. 19
7. 33
8. 120
9. 16
10. 65
11. 32
12. 45
13. 17

Math Masters

14. 15.5
15. 3.5
16. 8.5
17. 21.5
18. 26
19. 36
20. 4.6

Answers: Technique 11

Basic Exercises		Math Masters	
1. 180	8. 360	14. 225	
2. 320	9. 490	15. 385	
3. 230	10. 280	16. 195	
4. 120	11. 440	17. 290	
5. 390	12. 210	18. 460	
6. 260	13. 340	19. 3,300	
7. 420		20. 13,000	

Answers: Technique 12

Basic Exercises		Math Masters	
1. 8.6	8. 20.6	14. 6.8	
2. 14.2	9. 9.6	15. 5.7	
3. 11.2	10. 3.4	16. 14.6	
4. 4.4	11. 15.2	17. 4.44	
5. 17.4	12. 22.4	18. 27	
6. 7.2	13. 6.2	19. 7.8	
7. 18.8		20. 1.62	

Answers to Day 3 Review Quiz

Basic Exercises	Math Masters
1. 180	9. 140
2. 65	10. 9.5
3. 210	11. 440
4. 7.2	12. 2.24
5. 15.6	
6. 360	
7. 32	
8. 108	

Answers: Technique 13

Basic Exercises		Math Masters	
1. 275	8. 825	14. 253	
2. 473	9. 1,056	15. 102.3	
3. 891	10. 396	16. 5.28	
4. 363	11. 176	17. 869	
5. 682	12. 737	18. 45,100	
6. 649	13. 1,078	19. 60,500	
7. 308		20. 92.4	

Answers: Technique 14

Basic Exercises		Math Masters	
1. 4,225	8. 3,025	14. 225	
2. 1,225	9. 625	15. 122,500	
3. 7,225	10. 11,025	16. 72.25	
4. 225	11. 1,225	17. 3,025	
5. 9,025	12. 4,225	18. 0.9025	
6. 2,025	13. 13,225	19. 20,250	
7. 5,625		20. 422,500	

Answers: Technique 15

Basic Exercises		Math Masters	
1. 1,900	8. 700	14. 1,150	
2. 1,100	9. 1,400	15. 1,350	
3. 800	10. 600	16. 1,550	
4. 400	11. 1,600	17. 1,800	
5. 1,700	12. 1,200	18. 3,500	
6. 2,200	13. 2,400	19. 21.5	
7. 2,300		20. 210	

Answers: Technique 16

Basic Exercises		Math Masters	
1. 36	8. 22	14. 180	
2. 4.4	9. 25.6	15. 5.6	
3. 28	10. 12	16. 8.4	
4. 128	11. 280	17. 3.52	
5. 340	12. 68	18. 18.8	
6. 16.4	13. 32	19. 2.84	
7. 4.8		20. 7.6	

Answers to Day 4 Review Quiz

Basic Exercises	Math Masters
1. 396	9. 473
2. 7,225	10. 225
3. 1,600	11. 24
4. 4.8	12. 14.4
5. 28	
6. 800	
7. 2,025	
8. 935	

Answers: Technique 17

Basic Exercises		Math Masters
1. 98	8. 81	14. 6,800
2. 45	9. 583	15. 135
3. 51	10. 91	16. 34.1
4. 76	11. 57	17. 900
5. 72	12. 84	18. 72
6. 165	13. 48	19. 600
7. 52		20. 42

Answers: Technique 18

Basic Exercises		Math Masters
1. 4	8. 3	14. 4
2. 40	9. 4	15. 6
3. 60	10. 80	16. 6
4. 50	11. 14	17. 3
5. 40	12. 6	18. 17
6. 7	13. 7	19. 3
7. 90		20. 0.7

Answers: Technique 19

Basic Exercises		Math Masters
1. 3,721	8. 961	14. 82.81
2. 441	9. 6,561	15. 961
3. 2,601	10. 441	16. 65.61
4. 121	11. 3,721	17. 504,100
5. 8,281	12. 2,601	18. 441
6. 1,681	13. 1,681	19. 168.1
7. 5,041		20. 26.01

Answers: Technique 20

Basic Exercises		Math Masters
1. 399	8. 195	14. 122.4
2. 143	9. 624	15. 22.4
3. 2,024	10. 9,999	16. 48.99
4. 8,099	11. 255	17. 90.24
5. 168	12. 4,224	18. 168
6. 5,624	13. 6,399	19. 159.9
7. 2,499		20. 302.4

Answers to Day 5 Review Quiz

Basic Exercises

1. 84
2. 6
3. 961
4. 2,499
5. 600
6. 6,561
7. 6
8. 81

Math Masters

9. 72
10. 37.21
11. 6
12. 39.9

Midterm Exam Answers

Basic Exercises

1. 32
2. 138
3. 124
4. 114
5. 24
6. 37
7. 28
8. 57
9. 208
10. 55

11. 360
12. 4.4
13. 682
14. 1,225
15. 1,700
16. 12.8
17. 143
18. 14
19. 1,681
20. 899

Math Masters

21. 223
22. 190
23. 174
24. 77
25. 48
26. 140

27. 7.6
28. 242
29. 13,225
30. 6.8
31. 4
32. 224

Answers: Technique 21

Basic Exercises

1. 480
2. 210
3. 330
4. 540
5. 240
6. 420
7. 660

8. 390
9. 570
10. 960
11. 780
12. 510
13. 990

Math Masters

14. 690
15. 93
16. 1,080
17. 270
18. 12,000
19. 480
20. 126

Answers: Technique 22

Basic Exercises

1. 10,914
2. 11,881
3. 10,807
4. 10,816
5. 11,556
6. 11,009
7. 11,024

8. 10,302
9. 10,815
10. 11,340
11. 11,663
12. 10,201
13. 10,506

Math Masters

14. 12,075
15. 12,584
16. 12,099
17. 13,596
18. 11,872
19. 14,790
20. 19,998

Answers: Technique 23

Basic Exercises

1. 1,800
2. 900
3. 2,700
4. 3,300
5. 600
6. 2,100
7. 3,600

8. 4,200
9. 5,400
10. 1,200
11. 6,600
12. 5,100
13. 4,800

Math Masters

14. 900
15. 270
16. 330
17. 1,350
18. 18
19. 1,950
20. 1,050

Answers: Technique 24

Basic Exercises

1. 1.6
2. 2.8
3. 48
4. 3.6
5. 6
6. 5.2
7. 80

8. 44
9. 6.8
10. 13.2
11. 2.4
12. 56
13. 8.8

Math Masters

14. 96
15. 12.4
16. 7.2
17. 1.08
18. 116
19. 84
20. 16

Answers to Day 6 Review Quiz

Basic Exercises	Math Masters
1. 780	9. 480
2. 11,124	10. 13,566
3. 1,200	11. 180
4. 6	12. 8.4
5. 16	
6. 4,200	
7. 11,235	
8. 660	

Answers: Technique 25

Basic Exercises		Math Masters
1. 108	8. 153	14. 1,170
2. 171	9. 135	15. 423
3. 216	10. 252	16. 34.2
4. 126	11. 315	17. 3,060
5. 234	12. 207	18. 52.2
6. 144	13. 225	19. 801
7. 405		20. 603

Answers: Technique 26

Basic Exercises		Math Masters
1. 168	8. 204	14. 168
2. 396	9. 900	15. 2,280
3. 180	10. 192	16. 78
4. 420	11. 276	17. 102
5. 216	12. 300	18. 252
6. 264	13. 384	19. 540
7. 660		20. 4.08

Answers: Technique 27

Basic Exercises		Math Masters
1. 4,000	8. 10,000	14. 1,500
2. 8,000	9. 70,000	15. 4,500
3. 6,000	10. 3,000	16. 7,500
4. 9,000	11. 5,000	17. 300
5. 2,000	12. 1,000	18. 80
6. 11,000	13. 20,000	19. 400
7. 12,000		20. 1,100

Answers: Technique 28

Basic Exercises		Math Masters
1. 8.88	8. 12	14. 4.8
2. 24	9. 32	15. 3.2
3. 96	10. 17.6	16. 2.8
4. 16	11. 4	17. 88.8
5. 88	12. 6.4	18. 5.6
6. 48	13. 56	19. 16
7. 7.2		20. 40

Answers to Day 7 Review Quiz

Basic Exercises	Math Masters
1. 315	9. 21.6
2. 300	10. 540
3. 3,000	11. 200
4. 5.6	12. 48
5. 72	
6. 12,000	
7. 396	
8. 144	

Answers: Technique 29

Basic Exercises		Math Masters
1. 54	8. 5.6	14. 36
2. 12	9. 40	15. 16
3. 3.2	10. 24	16. 260
4. 48	11. 66	17. 14
5. 6.2	12. 4.6	18. 4.2
6. 8	13. 28	19. 6.4
7. 6		20. 4,400

Answers: Technique 30

Basic Exercises		Math Masters
1. 2,809	8. 3,249	14. 29.16
2. 3,364	9. 2,704	15. 2,704
3. 2,601	10. 3,364	16. 260,100
4. 3,481	11. 2,601	17. 34.81
5. 3,136	12. 2,809	18. 3,025
6. 2,916	13. 3,136	19. 32,490
7. 3,025		20. 28.09

Answers: Technique 31

Basic Exercises		Math Masters
1. 224	8. 190	14. 240
2. 415	9. 308	15. 770
3. 198	10. 558	16. 558
4. 188	11. 468	17. 2,070
5. 496	12. 348	18. 693
6. 324	13. 288	19. 990
7. 228		20. 3,150

Answers: Technique 32

Basic Exercises		Math Masters
1. 145	8. 594	14. 1,770
2. 406	9. 234	15. 1,470
3. 234	10. 345	16. 1,330
4. 384	11. 264	17. 725
5. 267	12. 441	18. 810
6. 272	13. 112	19. 1,185
7. 133		20. 1,188

Answers to Day 8 Review Quiz

Basic Exercises		Math Masters
1. 24	5. 114	9. 6.2
2. 2,916	6. 432	10. 32.49
3. 294	7. 3,364	11. 990
4. 294	8. 32	12. 570

Answers: Technique 33

Basic Exercises

1. 600   (actual answer = 594)
2. 900   (actual answer = 891)
3. 500   (actual answer = 510)
4. 1,600   (actual answer = 1,632)
5. 400   (actual answer = 396)
6. 700   (actual answer = 693)
7. 2,200   (actual answer = 2,244)
8. 1,900   (actual answer = 1,938)
9. 2,400   (actual answer = 2,376)
10. 1,500   (actual answer = 1,485)
11. 800   (actual answer = 816)
12. 3,300   (actual answer = 3,366)
13. 1,100   (actual answer = 1,089)

Math Masters

14. 3,100   (actual answer = 3,069)
15. 2,800   (actual answer = 2,856)
16. 250   (actual answer = 255)
17. 29   (actual answer = 28.71)
18. 1,300   (actual answer = 1,287)
19. 270   (actual answer = 275.4)
20. 180   (actual answer = 183.6)

Answers: Technique 34

Basic Exercises

1. 7.8   (actual answer = 7.8787 . . .)
2. 18   (actual answer = 18.1818 . . .)
3. 6.3   (actual answer ≈ 6.18)
4. 21   (actual answer ≈ 20.58)
5. 4.2   (actual answer = 4.2424 . . .)
6. 24   (actual answer = 24.2424 . . .)
7. 6.9   (actual answer ≈ 6.76)
8. 15   (actual answer ≈ 14.71)
9. 3.3   (actual answer = 3.333 . . .)
10. 12   (actual answer = 12.1212 . . .)
11. 4.8   (actual answer ≈ 4.71)
12. 7.5   (actual answer ≈ 7.35)
13. 22.5   (actual answer = 22.7272 . . .)

Math Masters

14. 84   (actual answer = 84.8484 . . .)
15. 36   (actual answer ≈ 35.29)
16. 60   (actual answer ≈ 58.82)
17. 1.56   (actual answer = 1.5757 . . .)
18. 9.3   (actual answer = 9.3939 . . .)
19. 3.9   (actual answer ≈ 3.82)
20. 5.4   (actual answer ≈ 5.29)

Answers: Technique 35

Basic Exercises

1. 1,700   (actual answer = 1,666)
2. 900   (actual answer = 882)
3. 1,300   (actual answer = 1,326)
4. 3,300   (actual answer = 3,366)
5. 3,600   (actual answer = 3,528)
6. 2,900   (actual answer = 2,842)
7. 650   (actual answer = 663)
8. 4,350   (actual answer = 4,437)
9. 1,450   (actual answer = 1,421)
10. 2,050   (actual answer = 2,009)
11. 4,200   (actual answer = 4,284)
12. 4,950   (actual answer = 5,049)
13. 1,650   (actual answer = 1,617)

Math Masters

14. 190   (actual answer = 186.2)
15. 70   (actual answer = 71.4)
16. 480   (actual answer = 489.6)
17. 380   (actual answer = 372.4)
18. 3,050   (actual answer = 2,989)
19. 120   (actual answer = 122.4)
20. 460   (actual answer = 469.2)

Answers: Technique 36

Basic Exercises

1. 6.6   (actual answer $\approx$ 6.73)
2. 1.9   (actual answer $\approx$ 1.94)
3. 12.8   (actual answer $\approx$ 12.55)
4. 8.2   (actual answer $\approx$ 8.04)
5. 3.2   (actual answer $\approx$ 3.27)
6. 14.4   (actual answer $\approx$ 14.69)
7. 5.4   (actual answer $\approx$ 5.29)
8. 13.8   (actual answer $\approx$ 13.53)
9. 7.6   (actual answer $\approx$ 7.76)
10. 16.8   (actual answer $\approx$ 17.14)
11. 3.8   (actual answer $\approx$ 3.73)
12. 1.74   (actual answer $\approx$ 1.71)
13. 18.2   (actual answer $\approx$ 18.57)

Math Masters

14. 11.2   (actual answer $\approx$ 11.43)
15. 2.8   (actual answer $\approx$ 2.75)
16. 44   (actual answer $\approx$ 43.14)
17. 8.8   (actual answer $\approx$ 8.98)
18. 15.2   (actual answer $\approx$ 15.51)
19. 4.8   (actual answer $\approx$ 4.71)
20. 7   (actual answer $\approx$ 6.86)

Answers to Day 9 Review Quiz

Basic Exercises

1. 800   (actual answer = 792)
2. 4.2   (actual answer ≈ 4.12)
3. 3,600   (actual answer = 3,672)
4. 5.4   (actual answer ≈ 5.51)
5. 7.6   (actual answer ≈ 7.45)
6. 3,300   (actual answer = 3,234)
7. 7.5   (actual answer = 7.5757 . . .)
8. 1,600   (actual answer = 1,632)

Math Masters

9. 280   (actual answer = 277.2)
10. 36   (actual answer ≈ 35.29)
11. 900   (actual answer = 882)
12. 15.2   (actual answer ≈ 14.90)

Answers: Technique 37

Basic Exercises

1. 2,800   (actual answer = 2,772)
2. 1,600   (actual answer = 1,584)
3. 2,400   (actual answer = 2,412)
4. 800   (actual answer = 804)
5. 3,000   (actual answer = 2,970)
6. 1,800   (actual answer = 1,782)
7. 1,400   (actual answer = 1,407)
8. 3,600   (actual answer = 3,618)
9. 4,000   (actual answer = 3,960)
10. 2,200   (actual answer = 2,178)
11. 4,800   (actual answer = 4,824)
12. 2,000   (actual answer = 2,010)
13. 1,000   (actual answer = 990)

Math Masters

14. 5,800   (actual answer = 5,742)
15. 5,200   (actual answer = 5,226)
16. 66   (actual answer = 66.33)
17. 320   (actual answer = 316.8)
18. 260   (actual answer = 257.4)
19. 640   (actual answer = 643.2)
20. 54   (actual answer = 54.27)

Answers: Technique 38

Basic Exercises

1. 4.8 (actual answer = 4.8484 . . .)
2. 4.2 (actual answer = 4.2424 . . .)
3. 2.4 (actual answer ≈ 2.39)
4. 6.6 (actual answer ≈ 6.57)
5. 7.8 (actual answer = 7.8787 . . .)
6. 5.4 (actual answer = 5.4545 . . .)
7. 27 (actual answer ≈ 26.87)
8. 45 (actual answer ≈ 44.78)
9. 18 (actual answer = 18.1818 . . .)
10. 105 (actual answer = 106.0606 . . .)
11. 6.9 (actual answer ≈ 6.87)
12. 9.6 (actual answer ≈ 9.55)
13. 8.7 (actual answer = 8.7878 . . .)

Math Masters

14. 5.1 (actual answer = 5.1515 . . .)
15. 12 (actual answer ≈ 11.94)
16. 750 (actual answer ≈ 746.26)
17. 3.9 (actual answer = 3.9393 . . .)
18. 7.2 (actual answer = 7.2727 . . .)
19. 8.4 (actual answer ≈ 8.36)
20. 108 (actual answer ≈ 107.46)

Answers: Technique 39

Basic Exercises

1. 35.2 (actual answer = 35.555 . . .)
2. 48.4 (actual answer = 48.888 . . .)
3. 2.75 (actual answer = 2.777 . . .)
4. 6.71 (actual answer = 6.777 . . .)
5. 62.7 (actual answer = 63.333 . . .)
6. 52.8 (actual answer = 53.333 . . .)
7. 7.04 (actual answer = 7.111 . . .)
8. 1.76 (actual answer = 1.777 . . .)
9. 44 (actual answer = 44.444 . . .)
10. 56.1 (actual answer = 56.666 . . .)
11. 9.35 (actual answer = 9.444 . . .)
12. 31.9 (actual answer = 32.222 . . .)
13. 16.5 (actual answer = 16.666 . . .)

Math Masters

14. 10.67 (actual answer = 10.777 . . .)
15. 12.1 (actual answer = 12.222 . . .)
16. 38.5 (actual answer = 38.888 . . .)
17. 8.36 (actual answer = 8.444 . . .)
18. 8.8 (actual answer = 8.888 . . .)
19. 1.54 (actual answer = 1.555 . . .)
20. 1.1 (actual answer = 1.111 . . .)

Answers: Technique 40

Basic Exercises

1. 36   (actual answer = 36.3636 . . .)
2. 12.6   (actual answer = 12.7272 . . .)
3. 54   (actual answer = 54.5454 . . .)
4. 9   (actual answer = 9.0909 . . .)
5. 14.4   (actual answer = 14.5454 . . .)
6. 20.7   (actual answer = 20.9090 . . .)
7. 32.4   (actual answer = 32.7272 . . .)
8. 63   (actual answer = 63.6363 . . .)
9. 72   (actual answer = 72.7272 . . .)
10. 13.5   (actual answer = 13.6363 . . .)
11. 81   (actual answer = 81.8181 . . .)
12. 18   (actual answer = 18.1818 . . .)
13. 40.5   (actual answer = 40.9090 . . .)

Math Masters

14. 21.6   (actual answer = 21.8181 . . .)
15. 27   (actual answer = 27.2727 . . .)
16. 50.4   (actual answer = 50.9090 . . .)
17. 117   (actual answer = 118.1818 . . .)
18. 31.5   (actual answer = 31.8181 . . .)
19. 6.75   (actual answer = 6.8181 . . .)
20. 80.1   (actual answer = 80.9090 . . .)

Answers to Day 10 Review Quiz

Basic Exercises

1. 4,800   (actual answer = 4,752)
2. 10.5   (actual answer ≈ 10.45)
3. 27.5   (actual answer = 27.777 . . .)
4. 270   (actual answer = 272.7272 . . .)
5. 13.5   (actual answer = 13.6363 . . .)
6. 4.84   (actual answer = 4.888 . . .)
7. 9.6   (actual answer = 9.6969 . . .)
8. 1,600   (actual answer = 1,608)

Math Masters

9. 26   (actual answer = 25.74)
10. 3.9   (actual answer ≈ 3.88)
11. 47.3   (actual answer = 47.777 . . .)
12. 36   (actual answer = 36.3636 . . .)

Answers to Final Exam

Basic Exercises

1. 37
2. 155
3. 126
4. 36
5. 54
6. 58
7. 45
8. 230
9. 737
10. 9,025
11. 36
12. 4
13. 540
14. 11,336
15. 6,600
16. 4.8
17. 126
18. 420
19. 9,000
20. 32
21. 28
22. 2,809
23. 324
24. 273
25. 700   (actual answer = 693)

26. 15   (actual answer ≈ 14.71)
27. 700   (actual answer = 686)
28. 3.2   (actual answer ≈ 3.14)
29. 1,800   (actual answer = 1,782)
30. 18   (actual answer ≈ 17.91)
31. 44   (actual answer = 44.444 . . .)
32. 12.6   (actual answer = 12.7272 . . .)

Math Masters

33. 204
34. 159
35. 204
36. 7.8
37. 6,800
38. 3.99
39. 270
40. 12,688
41. 80
42. 780
43. 24
44. 270,400
45. 378
46. 12   (actual answer = 12.1212 . . .)
47. 90   (actual answer = 88.2)
48. 140   (actual answer = 140.7)
49. 47.3   (actual answer = 47.777 . . .)
50. 8.1   (actual answer = 8.1818 . . .)

SUMMARY OF THE 40 EXCELL-ERATED MATH TECHNIQUES

NOTE: These descriptions are as brief as possible to enhance their usefulness. In some cases, it should be understood that zeroes and decimal points also play a part in the computation.

Technique 1:

Add numbers out of order, in combinations of 10, and in groups of two and three.

Technique 2:

Add from left to right, first the tens digit, then the ones.

Technique 3:

Add from left to right, first all the tens digits, then all the ones digits.

Technique 4:

Add by converting one number to a multiple of 10.

Technique 5:

Subtract by adding, working from left to right.

Technique 6:

Subtract numbers on opposite sides of 100 by adding the distances from 100.

Technique 7:

Subtract by converting the subtrahend to a multiple of 10.

Technique 8:

Subtract in two steps, first the tens digit, then the ones.

Technique 9:

Multiply by 4 by doubling twice.

Technique 10:

Divide by 4 by halving twice.

Technique 11:

Multiply by 5 by dividing by 2.

Technique 12:

Divide by 5 by multiplying by 2.

Technique 13:

Multiply by 11 by inserting the digits' sum in the middle.

Technique 14:

Square a number ending in 5 by multiplying the tens digit by the next whole number, and affixing the number 25.

Technique 15:

Multiply by 25 by dividing by 4.

Technique 16:

Divide by 25 by multiplying by 4.

Technique 17:

Multiply by 1.5, 2.5, etc., by doubling the 1.5, 2.5, etc., and halving the other number.

Technique 18:

Divide by 1.5, 2.5, etc., by doubling both the dividend and divisor.

Technique 19:

Square a number ending in 1, as in the following example: $21^2 = 20^2 + 20 + 21 = 441$.

Technique 20:

Multiply two numbers whose difference is 2 by squaring the number in the middle and subtracting 1.

Technique 21:

Multiply by 15 by multiplying by 10 and adding half the product.

Technique 22:

Multiply two numbers just over 100 by writing 1, then affixing the ones digits' sum, then affixing the ones digits' product.

Technique 23:

Multiply by 75 by multiplying by ¾.

Technique 24:

Divide by 75 by multiplying by 1⅓.

Technique 25:

Multiply by 9 by multiplying by 10 and then subtracting the number.

Technique 26:

Multiply by 12 by multiplying by 10 and then adding twice the number.

Technique 27:

Multiply by 125 by dividing by 8.

Technique 28:

Divide by 125 by multiplying by 8.

Technique 29:

Divide by 15 by multiplying by ⅔.

Technique 30:

Square a number beginning in 5 by adding 25 to the ones digit, and then affixing the square of the ones digit.

Technique 31:

Multiply by regrouping, as in the following example: $32 \times 7 = (30 \times 7) + (2 \times 7) = 224$.

Technique 32:

Multiply by augmenting, as in the following example: $39 \times 6 = (40 \times 6) - (1 \times 6) = 234$.

Technique 33:

Estimate multiplication by 33 or 34 by dividing by 3.

Technique 34:

Estimate division by 33 or 34 by multiplying by 3.

Technique 35:

Estimate multiplication by 49 or 51 by dividing by 2.

Technique 36:

Estimate division by 49 or 51 by multiplying by 2.

Technique 37:

Estimate multiplication by 66 or 67 by multiplying by ⅔.

Technique 38:

Estimate division by 66 or 67 by multiplying by 1.5.

Technique 39:

Estimate division by 9 by multiplying by 11.

Technique 40:

Estimate division by 11 by multiplying by 9.